NEW JERSEY ASK7

MATH TEST

JOHN T. NERAL, M.A.

BARRON'S

About the Author

John Neral has been teaching mathematics for 12 years and currently works as a district mathematics coordinator in Northern New Jersey. In addition, he serves as a consultant for various educational companies and has presented numerous workshops at the local, state, and national levels. John would like to thank his parents, family, friends, and colleagues who have been and continue to be a great source of strength and support.

© Copyright 2008 by Barron's Educational Series, Inc.

All inquiries should be addressed to:
Barron's Educational Series, Inc.
250 Wireless Blvd.
Hauppauge, NY 11788
www.barronseduc.com

ISBN-13: 978–0-7641–3943-7
ISBN-10: 0–7641-3943–6

Library of Congress Catalog No.: 2008003789

Library of Congress Cataloging-in-Publication Data

Neral, John T.
    NJ ask7 math test / by John T. Neral.
        p. cm.
    Includes bibliographical references and index.
    ISBN-13: 978-0-7641-3943-7 (alk. paper)
    ISBN-10: 0-7641-3943-6 (alk. paper)
    1. Mathematics—Examinations—Study guides. 2. Mathematics—Study and teaching (Middle school)—New Jersey. 3. Examinations—New Jersey—Study guides.   I. Title.

    QA43.N385 2008
    510.76—dc22

                                                                2008003789

Printed and bound in the United States of America
9 8 7 6 5 4 3 2 1

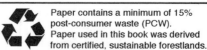

Paper contains a minimum of 15% post-consumer waste (PCW). Paper used in this book was derived from certified, sustainable forestlands.

CONTENTS

INTRODUCTION

It's time to get ready for the NJ ASK7 Math Assessment. The New Jersey Department of Education has established a series of grade-level assessments designed to determine how proficient you are in mathematics. The test might seem a little intimidating, but it is really an indicator to see how much you know and how good you are at solving math problems. At school, you have learned about many mathematics concepts and topics. Now, it's time to apply that information.

The NJ ASK7 Math Assessment tests your knowledge in four key areas: Number and Numerical Operations; Geometry and Measurement; Patterns and Algebra; and Data Analysis, Probability, and Discrete Mathematics. These areas are the same content areas determined by the National Council of Teachers of Mathematics Standards and the New Jersey Core Content Curriculum Standards. Throughout the test there are an equal number of questions from each area. Questions are either multiple-choice, short constructed response, or extended constructed response. You can use a calculator for the entire assessment.

In a multiple-choice question, four possible answers are given. You are given one (1) point for each correct answer. There is no penalty for an incorrect response. The short constructed response questions require you to solve a problem and provide an answer. You are given (1) point for each correct answer. The extended constructed response questions provide opportunities for you to write and explain your answer, construct a diagram or graph, and show calculations when directed. Each extended response question is worth up to three (3) points, which are determined based upon a rubric. The rubric is used by the examiner to determine the "correctness" of your answer. You can earn "partial credit" even if you do not get the entire question correct. Your raw score (number of points you have earned) is translated into an ASK score. An Advanced Proficient grade is earned when you score 250 or greater. A Proficient grade is earned if you score between 200 and 249.

Finally, a score below 200 is deemed Partially Proficient. Your school district may use these scores as criteria to evaluate how you are doing as well as to determine placement into future mathematics courses.

As you explore and navigate through the book, you will see that Chapters 1–4 focus on the main areas of the assessment. Chapter 5 focuses on various problem-solving and test-taking strategies that can be used throughout the test. Each chapter presents a series of lessons. Each lesson provides a description of the topic, several examples, and sample test questions. Additionally, there are two practice tests you can use to track your progress and knowledge of the material. Answers to all practice questions are provided.

Finally, the goal of this book is to help you better understand mathematics, become a better problem solver, help you gain confidence in taking the NJ ASK7 Math Assessment, and increase your knowledge base.

A LETTER TO THE STUDENT

Dear Student,

I am very excited that you are going to work with Barron's NJ ASK7 Math prep book. Taking tests is not easy. However, you have probably heard your teachers talk endlessly about the NJ ASK7 and being prepared for it. This test is very important, and your teachers, your parents/guardians, friends, and I want you to do very well on it.

As you look through the book, you will notice a series of chapters and lessons. Each lesson is designed with an explanation followed by some examples. After the examples, you will see there are sample test questions. The sample test questions—multiple-choice, short constructed response, and extended constructed response—are written to parallel the kind of format and wording you will see on the NJ ASK7 Math. You must know that there is no penalty for a wrong answer to a multiple-choice or short constructed response question and that you should always write something for the extended constructed response items. The extended constructed response items are scored using a rubric, and partial credit is given for some amount of correctness.

Math books are difficult to read, but I tried to write this book in such a way that the explanations would be easy to understand and the question references would appeal to everyone. You'll also notice VIPs throughout the book. A VIP is a Very Important Point and emphasizes a concept or strategy needed for the test. Refer to the VIPs often. They will be very helpful.

Most importantly, know that I hope you gain more confidence as a math student, increase your problem-solving skills, and do very well on the assessment.

Good Luck & Best Wishes,

John T. Neral

New Jersey Assessment of Skills and Knowledge
Grade 7
MATHEMATICS REFERENCE SHEET*

Use the information below to answer questions on the Mathematics section of the
Grade Seven Assessment of Skills and Knowledge (NJ ASK 7).

The sum of the measures of the interior angles of a triangle = 180°

Distance = rate $\times$ time

Simple Interest Formula: $A = p + prt$ Compound Interest Formula : $A = p(1 + r)^t$

A = amount after t years; p = principal; r = annual interest rate; t = number of years

---

$\pi \approx 3.14$ or $\frac{22}{7}$

Square
Area = s^2
Perimeter = $4s$

Rectangle
Area = lw
Perimeter = $2l + 2w$

Circle
Area = πr^2
Circumference = $2\pi r$
= πd

Triangle
Area = $\frac{1}{2} bh$

Parallelogram
Area = bh

Trapezoid
Area = $\frac{1}{2} h(b_1 + b_2)$

Rectangular Prism
Volume = lwh
Surface Area =
$2\,lw + 2\,wh + 2\,lh$
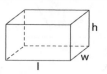

Cylinder
Volume = $\pi r^2 h$
Surface Area =
$2\,\pi rh + 2\pi r^2$

Cone
Volume = $\frac{1}{3} \pi r^2 h$

Pyramid
Volume = $\frac{1}{3} lwh$
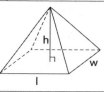

USE THE FOLLOWING EQUIVALENTS FOR YOUR CALCULATIONS

60 seconds = 1 minute	12 inches = 1 foot	10 millimeters = 1 centimeter
60 minutes = 1 hour	3 feet = 1 yard	100 centimeters = 1 meter
24 hours = 1 day	36 inches = 1 yard	10 decimeters = 1 meter
7 days = 1 week	5,280 feet = 1 mile	1000 meters = 1 kilometer
12 months = 1 year	1,760 yards = 1 mile	
365 days = 1 year		

8 fluid ounces = 1 cup	16 ounces = 1 pound
2 cups = 1 pint	2,000 pounds = 1 ton
2 pints = 1 quart	
4 quarts = 1 gallon	1000 milligrams = 1 gram
	100 centigrams = 1 gram
1000 milliliters (mL) = 1 liter (L)	10 grams = 1 dekagram
	1000 grams = 1 kilogram

9-13129

*Source: http://www.nj.gov/education/assessment/ms/5-7/gr7_NJASK06_math_ref.pdf

NJ ASK
Mathematics Manipulatives Sheet
Shapes, Protractor, and Ruler
Grades 5, 6, and 7 Only*

Regular Triangles

Note: For reference only— not drawn to scale.

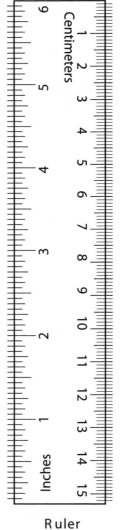

Ruler

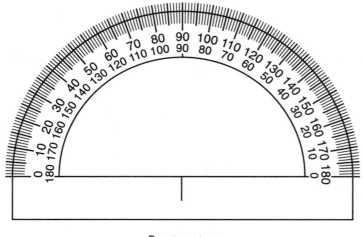

Protractor

*Source: http://www.state.nj.us/education/assessment/ms/5-7/2006-07ManipulativeSheet.pdf

<table>
<tr><td>Chapter</td></tr>
<tr><td>1</td></tr>
</table>

NUMBER AND NUMERICAL OPERATIONS

LESSON 1: INTEGERS

Integers are defined as positive and negative whole numbers. While we may not always refer to these numbers as integers, they are most commonly seen as the numbers which appear on a number line. Since zero is considered a whole number, it is also considered an integer.

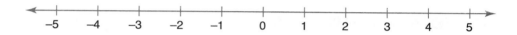

TO ADD TWO INTEGERS

- If you are adding two positive integers, the sum is always positive.
  Example: $5 + 3 = 8$

- If you are adding two negative integers, the sum will always be negative.
  Example: $(-3) + (-7) = (-10)$

- However, if you are adding one positive integer and one negative integer, the answer can be either positive or negative.
  Example: $3 + (-5) = ?$

One way to solve this problem is to draw it on a number line. The first integer becomes your starting point and the second integer tells how many places and which direction you would move on the number line. If the integer is positive, you move to the right. If the integer is negative, you move to the left.

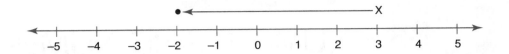

In this example, the X denotes the starting position, which is the first integer in the problem, 3. Then, we move five places to the left because the second integer is (–5). The number we land at is our sum, which is (–2).

Another way to solve this problem is to think of the problem in terms of money. Earning, finding, or having money can be considered positive, and spending, losing, or owing money can be considered negative. If you think of a story that models the problem, it can make it much easier to solve.

Given the expression 3 + (–5), we could solve the problem in this way. "If I have three dollars and I owe you five dollars, I don't have enough money to pay you back. If I give you my three dollars, how much money will I owe you?" Once I give you my three dollars, I still owe you two. Since owing money is considered a representation of negative numbers, my answer is (–2).

TO SUBTRACT INTEGERS

We can subtract integers very easily if we remember these three key words:

KEEP–CHANGE–CHANGE

Following this model, we will KEEP the first integer, CHANGE the subtraction sign to an addition sign, and then CHANGE the second integer to its opposite. Let's take a look at this example:

7 – 12 = ?

One way to solve the problem is to use the phrase KEEP–CHANGE–CHANGE. What that means is you will keep the first integer, 7, change the subtraction sign to addition, and then change the last integer, 12, to its opposite, (–12). This makes the problem 7 + (–12) and its solution is (–5).

Another method is to use the following organizer. Here, the headings of each column are listed, KEEP–CHANGE–CHANGE, and the problem 7 – 12 is placed accordingly. As already demonstrated, you will keep the first integer, change the subtraction sign to addition, and change the last integer to its opposite.

KEEP	CHANGE	CHANGE
7	–	12
7	+	(–12)

Now, our problem becomes 7 + (–12) and the answer is (–5). This organizer can be helpful for those students who need to use a graphic model to help organize his/her thoughts better.

Use the Table to review rules for multiplying and dividing integers.

Multiplying and Dividing Integers	Examples
When you multiply or divide two integers and both numbers are positive, the answer will be positive.	$8 \times 4 = 32$ $36 \div 4 = 9$
When you multiply or divide two integers and both numbers are negative, the answer will be positive.	$(-9) \times (-3) = 27$ $(-48) \div (-12) = 4$
When you multiply or divide two integers and one number is positive and the other is negative, the answer will be negative.	$7 \times (-5) = (-35)$ $32 \div (-8) = (-4)$
When you multiply or divide more than two integers, simply multiply or divide in order from left to right.	$4 \times (-3) \times (-2) \times 5 = 120$ $(-20) \cdot 5 \div (-2) = 50$

SAMPLE TEST QUESTIONS (For answers, see page 173.)

Multiple-Choice (Circle the correct answer.)

1. Which pair of integers has a sum of −9?

   A. −5 + 4

   B. −7 + −2

   C. −8 + 1

   D. −6 + 3

2. I am an integer that is 5 more than −3. Who am I?

   A. −8

   B. −2

   C. 2

   D. 8

3. We are two integers that have a product of 15 and have a sum of −8. Who are we?

   A. 3 and 5

   B. −3 and 5

   C. 3 and −5

   D. −3 and −5

4. During a recent Philadelphia Eagles football game, the Eagles gained 5 yards on the first down, lost 7 yards on the second down, and gained 11 yards on the third down. In three downs, they

   A. gained more than 10 yards.

   B. gained less than 10 yards.

   C. gained exactly 10 yards.

   D. gained exactly 18 yards.

5. Jake's Office Supply Company recorded $250,000 in profits last year. This year, the company lost $180,000. Which integer expresses the company's profit/loss over these two years?

    A. –$430,000

    B. –$70,000

    C. +$70,000

    D. +$430,000

Short Constructed Response

6. What is $(-4) + 7 + (-8) + 11 + (-5)$? _____

7. On Monday, Grace earned $28.00 at work. She spent $4.00 on lunch and gave $9.00 to her younger sister.

    How much money does she have left? _____

Extended Constructed Response (Use the space provided to show your work.)

3	1	(–2)
2	0	(–1)
(–3)	4	(–4)

At a local fair, you decide to play a ball toss game where you toss balls into a 9-square box. You are given three balls. Assume that all balls land in the box. No more than one ball can land on a given number.

- What is the highest score with three balls?

- What is the lowest score with three balls?

- Is it possible to score 0? If so, how could you do it?

- If all three balls form a tic-tac-toe (three balls in a row either horizontally, vertically, or diagonally), what is the highest possible score from a tic-tac-toe?

LESSON 2: FRACTIONS, DECIMALS, AND PERCENTS

One of the great things about our number system is that we can express the same number in a variety of ways. This concept is known as equivalence. Typically, when we want to express a number that is greater than 0 but less than 1, we use a fraction, a decimal, or in some cases, a percent. Being able to determine equivalence for these numbers is an important concept in mathematics.

For example, in this picture, the shaded area (as shown by the shading) represents $\frac{1}{4}$.

We can say that $\frac{1}{4}$ of the box is shaded because only 1 out of the 4 boxes shown is shaded. If we were to express $\frac{1}{4}$ as a decimal, we would say that the shaded region is 0.25. And if we wanted to express $\frac{1}{4}$ as a percent, we would say that 25% of the entire figure is shaded.

Think of a fraction as a representation between the number of items in the part versus the number of items in the whole. The number on top, also known as the numerator, refers to the amount of items in the part, while the number on the bottom, also known as the denominator, refers to the amount of items in the whole. A proper fraction is a fraction whose numerator is smaller than its denominator. An improper fraction is when the numerator is larger than the denominator.

V.I.P. (Very Important Point)	
$\dfrac{\text{Numerator}}{\text{Denominator}}$	Always express fractions in simplest form. To express a fraction in simplest form, determine the greatest common factor of the numerator *and* denominator and divide by each.

Example

Simplify $\dfrac{20}{44}$ The greatest common factor of 12 and 44 is 4. So, divide both numbers by 4. 20 ÷ 4 = 5. 44 ÷ 4 = 11. Therefore, the fraction is simplified to equal $\dfrac{5}{11}$.

We should be able to express any number as a fraction, decimal, or percent. We should also be able to determine a rule or method for converting between these forms.

To convert a fraction to a decimal, you can simply divide the numerator by the denominator to obtain the decimal equivalent of that fraction.

Example

$$\frac{3}{8} = 0.375 \qquad\qquad \frac{1}{7} = 0.\overline{142857}$$

When converting to a decimal form, you will have either a terminating or a repeating decimal. Whenever you have a repeating decimal, make certain to use the correct bar notation showing which numbers are being repeated in the decimal form.

To convert a decimal to a percent, simply multiply the decimal by 100 or move the decimal point two places to the right.

Example (continues on page 8)

0.35 = 35%   0.774 = 77.4%

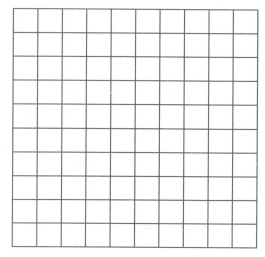

This is a 100-grid square. Since percents represent information out of 100, each square in this grid would represent 1%. For example, if you stood at the free-throw line on a basketball court and took 100 shots, how many do you think you would make? Let's say that Jessica has been practicing for an upcoming tournament, and each day she takes 100 shots in her backyard. Her father has been helping her train, and so he uses this 100-grid box to record when she makes a basket. Each time she does, he places an X in the box. The tournament is this Saturday. The day before, she practices one last time. Her results are shown in the following grid. Remember that each X represents a basket.

X	X	X	X	X	X		X	X	X
X	X	X	X	X	X	X	X	X	X
X	X	X	X	X	X	X	X	X	
X	X	X	X	X	X	X	X	X	X
X	X	X	X	X	X	X		X	X
X	X	X	X		X	X	X	X	X
X	X	X	X	X	X	X		X	X
	X	X	X	X	X	X	X	X	
X	X	X	X		X	X	X	X	X
X	X	X	X	X	X	X	X	X	X

How many baskets did Jessica make? Looking at the grid, we can see that she made 92 out of 100 baskets. This would equal 92% of the shots made. However, when Jessica goes to the tournament, she will only be allowed 25 shots in the first round. If her percentage remains the same, how many baskets should Jessica expect to make out of 25?

Here is a 25-grid box.

To determine this amount, you would need to take 92% of 25. We can calculate this a few different ways.

First, we can convert the percent to a decimal and multiply by 25.

$$0.92 \times 25 = 23$$

We could also solve this problem using a proportion. A proportion is a statement of equality between two ratios. A ratio describes a relationship between two items. Here, we could make a ratio between the number of baskets Jessica makes in relation to the total number of shots given. The other ratio would be the percent made over 100.

Perhaps an easier way might be to set up the Percent Proportion. The Percent Proportion is written as

$$\frac{is}{of} = \frac{\%}{100}$$

"Is" represents what the problem equals or the part or piece of the whole. The number placed in the "of" position is the number written after the word "of" in the problem and is referred to as the whole. The "$\frac{\%}{100}$" represent the percent. We can use the Percent Proportion to solve the question: What is 92% of 25?

$$\frac{is}{of} = \frac{\%}{100}$$

$$\frac{n}{25} = \frac{92}{100}$$
$$100n = 92 \cdot 25$$
$$100n = 2,300$$
$$\frac{100n}{100} = \frac{2,300}{100}$$
$$n = 23$$

To solve for n, you would multiply $92 \cdot 25$ and set that equal to $n \cdot 100$. Then, $2,300 = 100n$. Divide both sides by 100.

$$2300 \div 100 = 100n \div 100$$
$$23 = n$$

Percents help us express a relationship, and we can use that percent to transfer information to similar situations. In this example, even though Jessica is not going to have 100 shots at the free-throw tournament, if her percentage stays the same, then it is expected she will complete that same percentage of free throws regardless of how many shots she takes.

The table below shows the rules for determining equivalence between fractions, decimals, and percents. Having an understanding of these rules will help you to easily convert between these forms of numbers.

Conversion	Explanation
Decimal to fraction	Place the decimal over its place value. Example: $0.29 = \dfrac{29}{100}$
Decimal to percent	Move the decimal point two places to the right (same as multiplying by 100) and add the percent sign. Example: $0.29 = 29\%$
Fraction to decimal	Divide the numerator by the denominator. Example: $\dfrac{2}{5} = 2 \div 5 = 0.4$
Fraction to percent	Two Methods: 1. Convert fraction to a decimal and then convert that decimal to a percent. Example: $$\dfrac{2}{5} = 2 \div 5 = 0.4$$ $$0.4 \times 100 = 40\%$$ or 2. Use a proportion to determine the percent. Example: $$\dfrac{2}{5} = \dfrac{x}{100}$$ $$5x = 2 \cdot 100$$ $$5x = 200$$ $$\dfrac{5x}{5} = \dfrac{200}{5}$$ $$x = 40$$

Conversion	Explanation
Percent to decimal	Remove the percent sign and move the decimal point two places to the left (divide by 100). Example: 75% = 0.75
Percent to fraction	Remove the percent sign and place the value over 100. Simplify to lowest terms if necessary. Example: $75\% = \dfrac{75}{100} = \dfrac{3}{4}$

SAMPLE TEST QUESTIONS (For answers, see page 174.)

Multiple-Choice (Circle the correct answer.)

1. What is the equivalent form of $\dfrac{15}{24}$?

   A. 62.5

   B. 6.25

   C. 0.625

   D. 0.0625

2. Martha read 120 pages in her book, which has a total of 200 pages. Which fraction best represents the portion of the book that she has NOT read?

   A. $\dfrac{2}{5}$

   B. $\dfrac{3}{5}$

   C. $\dfrac{12}{20}$

   D. $\dfrac{1}{5}$

3. What percent best represents the portion shaded?

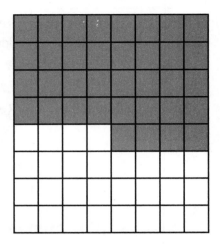

A. 25%

B. 46.875%

C. 56.25%

D. 60%

4. Yankee Stadium can seat 57,545 people. At a recent game, it was reported that approximately 97% of the tickets were sold. Based on that report, what is the best estimate for how many people attended the game?

A. 57,448

B. 55,819

C. 54,093

D. 1,726

5. Which set of numbers does NOT contain three numbers that are equivalent?

   A. $\frac{1}{2}$, 0.5, 50%

   B. $\frac{1}{5}$, 0.02, 20%

   C. $\frac{1}{8}$, 0.125, 12.5%

   D. $\frac{1}{20}$, 0.05, 5%

6. The New Jersey state lottery is $173,000,000. If you and your family win the lottery and receive 5% of that amount for the next twenty years, how much money would you receive each year?

   A. $172,999,995

   B. $164,350,000

   C. $34,600,000

   D. $8,650,000

7. There are 5 red marbles, 6 blue marbles, 3 green marbles, and 2 yellow marbles in a bag. What amount represents the number of blue and green marbles?

   A. 43.75%

   B. 50%

   C. $\frac{9}{16}$

   D. $\frac{11}{16}$

8. On a recent math test, Monique earned an 85%. Which value represents the amount Monique did NOT answer correctly?

   A. 3 out of 20

   B. 15

   C. $\dfrac{6}{40}$

   D. $\dfrac{7}{50}$

Short Constructed Response

9. A professional baseball player has a batting average of 0.325.

   Express the player's batting average as a fraction in simplest form. _____

10. During a recent free-throw basketball tournament, Jessica makes 27 out of 40 free throws. Express this amount as a decimal. _____

Extended Constructed Response (Use the space provided to show your work.)

Using the number line below find three different numbers located between *A* and *B*. Express one number as a fraction, another number as a percent, and another number as a decimal.

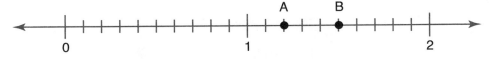

LESSON 3: WHOLE NUMBERS AND EXPONENTS

A number written in exponential form has a base and an exponent. The base is the number written with a larger font, and the exponent is the number written above the base using a smaller font. In the example, 5^4, the 5 is the base, and the 4 is the exponent.

By definition, the exponent tells us how many times to use the base as a factor. When we do that, we get a product of factors. When we evaluate the product of factors, we get a number in standard form or what the exponential form equals.

$$5^4 \qquad = \qquad 5 \times 5 \times 5 \times 5 \qquad = \qquad 625$$

| Exponential form | Product of factors | Standard form |

Examine this table to see how when we increase a base's exponent, we increase the number in standard form.

Exponential Form	Product of Factors	Standard Form
3^0		1
3^1	3	3
3^2	3×3	9
3^3	$3 \times 3 \times 3$	27
3^4	$3 \times 3 \times 3 \times 3$	81
3^5	$3 \times 3 \times 3 \times 3 \times 3$	243

V.I.P. (Very Important Points)

- Any number raised to the zero power will always equal 1.
- Whenever a number is raised to the second power, it is commonly referred to as being squared.
- Whenever a number is raised to the third power, it is commonly referred to as being cubed.

If you are using a scientific calculator to evaluate the standard form of a number written in exponential form, you need to use a key that looks like an up arrow (also known as a caret), ^, or a key that looks like this, y^x. To correctly use this feature, input the base first, press the exponent key on your calculator, and then enter the number for the exponent followed by the equal sign, enter key or EXE key, depending upon which calculator you are using. This will give you the standard form for any number given in exponential form.

SAMPLE TEST QUESTIONS (For answers, see pages 174–175.)

Multiple-Choice (Circle the correct answer.)

1. Which expression best represents $9 \times 9 \times 9 \times 9$?

   A. 9^4

   B. 4^9

   C. $(9)(4)$

   D. 9^9

2. Evaluate $5^4 + 5^2$.

   A. 650

   B. 657

   C. 1,049

   D. 15,625

3. Which expression is equivalent to 12^3?

   A. 12×3

   B. $12 \times 12 \times 12$

   C. 3^{12}

   D. $12 \times 12 \times 12 \times 12$

4. Which of these statements is FALSE?

   A. $2^3 = 8$

   B. $5^2 = 25$

   C. $4^0 = 0$

   D. $7^4 = 2,401$

5. If you are given three pennies on day 1 and each day that amount triples, how many pennies would you be given on day 5?

    A. 27

    B. 81

    C. 243

    D. 729

6. A certain bacteria doubles every 30 minutes. If there are 500 bacteria cells at 10:00 A.M., how many bacteria cells will there be at 12:00 noon?

    A. 16,000

    B. 8,000

    C. 4,000

    D. 2,000

7. $8^? = 32,768$

    A. 4

    B. 5

    C. 6

    D. 7

8. Which of the following is different than the others?

    A. 4^2

    B. 8^2

    C. 2^4

    D. 16^1

Short Constructed Response

9. Which number will make the statement true?

   $?^3 = 125$ _____

Extended Constructed Response (Use the space provided to show your work.)

On a recent math test, Brad says that $4^2 + 3^4 = 2^4 + 4^3$. Using complete sentences, evaluate Brad's answer, and if incorrect, explain the correct solution.

LESSON 4: COMPARING AND ORDERING NUMBERS

As we have stated before, numbers can be expressed in a variety of ways. When making comparisons between those numbers, you must find a common way to order them. When comparing integers, the larger number is the number that appears furthest to the right on the number line.

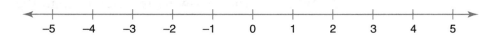

However, if you are comparing only negative numbers, the larger negative number is the one that is closest to zero.

When comparing percents, decimals, and fractions, it is best to express them in one form. For some, it might be easiest to express them as decimals and then compare. For others, it might be easiest to express them as fractions. We discussed how to convert between fractions, decimals, and percents in Lesson 2. So, whichever way is easiest for you, convert the numbers to that one form and make your comparisons.

V.I.P. (Very Important Point)

Always make certain to read the problems very carefully. Sometimes, students make mistakes because they did not read the problem correctly. Pay particular attention to how the numbers are to be ordered. For example, sometimes the question is worded to order the numbers from GREATEST to LEAST as opposed to being ordered from LEAST to GREATEST. Also, when writing the numbers in order, it may be helpful for you to cross off each number as you write it in your answer. This way, you will be certain not to miss any numbers! Try this strategy with some of the practice problems.

When making comparisons, we will use these inequality symbols.

Symbol	Meaning
<	Less than
>	Greater than
≤	Less than or equal to
≥	Greater than or equal to
=	Equal to
≠	Not equal to

SAMPLE TEST QUESTIONS (For answers, see pages 175–176.)

Multiple-Choice (Circle the correct answer.)

1. The heights of four adults are shown. Which one is the shortest?

   $5\frac{7}{12}$ feet, 68 inches, 5.5 feet, $5\frac{2}{3}$ feet

   A. $5\frac{7}{12}$ feet

   B. 68 inches

   C. 5.5 feet

   D. $5\frac{2}{3}$ feet

2. Four students ran for student council president. The table shows the portion of votes received by each candidate.

Candidate	Votes (Total)
Brad	0.28
Deanna	15%
Luke	$\frac{1}{4}$
Lindsay	$\frac{6}{25}$

Which candidate received the MOST votes?

A. Brad

B. Deanna

C. Luke

D. Lindsay

3. The guidance counselor at your school surveys five students to see how many hours they spent working on their Seventh Grade Science Fair Project. The responses were $3\frac{1}{4}$ hours, $2\frac{1}{2}$ hours, $3\frac{1}{3}$ hours, $3\frac{1}{5}$ hours, $2\frac{3}{4}$ hours. When ranked in order from least to greatest, which time is ranked third?

A. $2\frac{1}{2}$ hours

B. $2\frac{3}{4}$ hours

C. $3\frac{1}{5}$ hours

D. $3\frac{1}{4}$ hours

4.

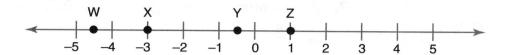

Looking at the number line, which number is represented by the letter *Y*?

A. $\dfrac{1}{2}$

B. $-\dfrac{1}{2}$

C. -1

D. -4.5

5. Which of the following numbers is the greatest?

A. 85%

B. $\dfrac{8}{11}$

C. $\dfrac{9}{10}$

D. 0.87

6. Which symbol makes the statement true? $\dfrac{14}{15}$ _____ $\dfrac{14}{16}$

A. $<$

B. $\leq$

C. $>$

D. $=$

Short Constructed Response

7. Write a number that is greater than 0.77 and less
than $\frac{4}{5}$.   _____

Extended Constructed Response (Use the space provided to show your work.)

1. Which is greater: 25% of 60 or 60% of 25?

   Determine which is greater and explain your reasoning using complete sentences.

2. Draw a number line showing two different numbers between $\frac{7}{10}$ and 0.85.

   Make certain to clearly label all points.

LESSON 5: ORDER OF OPERATIONS

There is an order of operations that we must follow whenever we solve a numerical expression. Perhaps you remember the phrase, "Please Excuse My Dear Aunt Sally" as a way of remembering the order of operations. Let's think of the order of operations in this way:

Please	Parentheses	This is a grouping symbol, and whenever you see that symbol you are to evaluate whatever is inside of it first. It's a way of saying that this is the most important part of the expression and MUST be evaluated first.

Excuse	Exponents	This is the second thing to be evaluated. Always be careful when evaluating exponents and follow your rules.
My Dear	**M**ultiplication or **D**ivision	This is the third step in the order of operations. Multiplication and division are performed in order from left to right. After you evaluate the parentheses and exponents, evaluate any multiplication or division problem in order from left to right.
Aunt **S**ally	**A**ddition or **S**ubtraction	This is the last step in the order of operations. Solve addition and subtraction problems in order from left to right after you have solved parentheses, exponents, and any multiplication and division problems.

It's important to know which operation you must solve first as well as being able to determine the correct value of the expression. Examine the following examples.

Examples

Solve the following expression using the order of operations:

$$5 + (9 - 2) \times 7 + 3 \qquad \text{Evaluate the parentheses first.}$$
$$5 + 7 \times 7 + 3 \qquad \text{Multiply } 7 \times 7.$$
$$5 + 49 + 3 \qquad \text{Add } 5 + 49.$$
$$54 + 3 \qquad \text{Add } 54 + 3.$$
$$57$$

Solve the following expression using the order of operations:

$3^4 - 2^4 + (15 - 3^2) + 27$	Evaluate the base and exponent inside of the parentheses.
$3^4 - 2^4 + (15 - 9) + 27$	Evaluate the parentheses.
$3^4 - 2^4 + 6 + 27$	Evaluate the exponents.
$81 - 16 + 6 + 27$	Subtract $81 - 16$.
$65 + 6 + 27$	Add $65 + 6$.
$71 + 27$	Add $71 + 27$.
98	

SAMPLE TEST QUESTIONS (For answers, see pages 176-177.)

Multiple-Choice (Circle the correct answer.)

1. What is $17 + 8 \times 3 - 5$?

    A. 70

    B. 36

    C. 26

    D. −38

2. Evaluate: $45 \div 9 + 6$

    A. 11

    B. 3

    C. 42

    D. 30

3. What is the answer to the first operation you would perform in the following problem?

    $$8 + 5 \times 2 + 7 - 4$$

    A. 13

    B. 10

    C. 9

    D. 3

4. In the expression $11 - 4 + 8 \div 2 \times 3$, what operation do you perform SECOND?

    A. Addition

    B. Subtraction

    C. Multiplication

    D. Division

5. Evaluate: $3^4 \div 9 + 5 \times 4$

    A. 92

    B. 56

    C. 52

    D. 29

6. In the expression $34 - 8 \div 13 \times (12 + 3)$, what is the operation you would perform first?

    A. Addition

    B. Subtraction

    C. Multiplication

    D. Division

7. What is the value of the expression $4 + 7 \times 9 - 5 + 12$?

    A. 74

    B. 82

    C. 106

    D. 507

Short Constructed Response

8. What is the value of $7^2 - 12 + 33 + 9 \times 2$?   _____

Extended Constructed Response (Use the space provided to show your work.)

Kamika solved the following problem. Make any corrections necessary, line by line, next to the problem.

$$13 - 9 + 18 \times 3 - 7^0 \times 8$$

$$13 - 9 + 18 \times 3 - 0 \times 8$$

$$4 + 18 \times 3 - 0 \times 8$$

$$22 \times 3 - 0 \times 8$$

$$66 - 0 \times 8$$

$$66 \times 8$$

$$528$$

LESSON 6: ABSOLUTE VALUE

Let's go back and review the number line we saw earlier in the chapter.

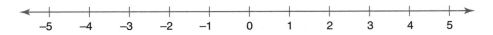

Consider zero as the center of the number line. Any nonzero number is a certain distance away from zero and that distance is what we call **absolute value**. Imagine a set of twins where one person is standing on +1 and the other twin is standing on (−1). What do they have in common? Each person is one unit away from zero. Now, if each twin took one step so they now stood on +2 and (−2), again they would have the same absolute value because they would now be two units from zero on the number line.

The absolute value symbol $|n|$ is used to tell you to determine how far that number is from zero. You can think of an absolute value symbol like a set of parentheses. Remember to solve whatever is inside of the absolute value symbol first before performing any additional calculations. For example:

$$|3| = 3 \qquad |-4| = 4 \qquad |2 - 7| + 3 = |-5| + 3 = 5 + 3 = 8$$

Not only can you evaluate the distance a single number is from zero, but you can also simplify an expression and take its absolute value. This distance is always a positive number because the distance is measured in units greater than zero. However, when solving absolute value problems, you can have answers that are negative.

How would you evaluate $-|-8|$

First, you would evaluate the absolute value of –8, which is 8. Then, since there is a negative sign outside of the absolute value symbol, you would take the opposite of 8 and make it (–8).

V.I.P. (Very Important Point)

Remember that the absolute value symbol is like a set of parentheses. Evaluate what is inside of them first, and then solve the rest of the problem.

Here are some other examples where the negative sign is seen outside of the absolute value symbol.

$$-|3| = -3 \qquad -|-7| = -7 \qquad -|-3-7| = -|-10| = -10$$

SAMPLE TEST QUESTIONS (For answers, see pages 177-178.)

Multiple-Choice (Circle the correct answer.)

1. Which number(s) have an absolute value of 7?

   A. 7

   B. –7

   C. 7 and –7

   D. None of the above

2. Which number could be a solution to $|x| - 5 = 9$?

   A. –14

   B. –4

   C. 4

   D. 0

3. Which relationship is TRUE?

    A. $|-3| < |1|$
    B. $|7 - 4| = |-1 + 4|$
    C. $-|-3| \neq |3|$
    D. $|9 - 4| < |-4 - 2|$

4. Which expression is equal to 2?

    A. $\left|2^0 + 1^3\right|$
    B. $-2 \cdot 2$
    C. $|-3| - |-2|$
    D. $\dfrac{|-8|}{-4}$

5. Given the problem $|x|$ = ?, the absolute value of any real number represented by x

    A. cannot be determined based on the information given.
    B. can be either negative or positive.
    C. is always negative.
    D. is always positive.

6. Which number(s) are 7 units away from 2?

    A. $|-5|$
    B. $|9|$
    C. $|-5|$ and $|-6 - 3|$
    D. $|-5|$ and $|10 - 1|$

Short Constructed Response

7. What is the value of x?

    $|-5| - x = 9$

    _____

Extended Constructed Response (Use the space provided to show your work.)

On the number line below, use an X to show the number(s) with an absolute value of 2.

Explain how you obtained your answer using complete sentences.

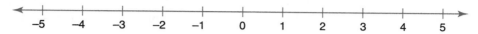

LESSON 7: RATIO AND PROPORTION

How many boys and how many girls are there in your class? How many students in your class have pets and how many do not? How many people in your class are wearing sneakers and how many are wearing shoes?

When answering these questions, you have to make a comparison about the two groups. For example, if there are 11 boys and 14 girls in your class, you could express the information as a ratio as $\frac{11}{14}$, 11 out of 14, or 11:14.

A ratio is a comparison of two numbers by division. When expressing information as a ratio, it is important to remain consistent in how you write that information. For instance, when asked how many boys and how many girls are in your class, if there are 11 boys and 14 girls, you would write the ratio as $\frac{11}{14}$, 11 out of

14, or 11:14. The number of boys MUST come first because that is how it was expressed in the question. It would be incorrect to put the number of girls before the number of boys. This is similar to writing an analogy. If the analogy is hot is to cold as summer is to winter, the relationship is reasonable because it is hot in the summer and cold in the winter.

Always express a ratio in simplest form. For example, if 8 students in your class are wearing sneakers and 12 students are wearing shoes, the ratio would be expressed as $\frac{8}{12}$. However, always express a ratio in simplest form. Therefore, $\frac{8}{12}$ would be written as $\frac{2}{3}$.

When two ratios are equal, we say that those ratios are in a proportion. A proportion is defined as two equal ratios whose cross products are equal. When proving that two ratios are equal, you can find the cross products by multiplying the top number of one ratio with the bottom number of the other ratio. Let's examine some examples.

When setting up a proportion, it is important to keep the same labels or titles in each part of the proportion. The organizer that appears later in the section may prove to be very helpful in setting up the proportion.

Example 1

Are the following two ratios in a proportion?

$$\frac{3}{8} = \frac{9}{24}$$

Solution: To solve the problem, determine if the cross products are equal.

$$\frac{3}{8} = \frac{9}{24}$$
$$3 \cdot 24 = 8 \cdot 9$$
$$72 = 72$$

Yes, each product is 72, and since the cross products are equal, the two ratios are equal and in proportion with each other.

Example 2

Let's say that your class takes an informal survey to see who likes Hollister and who likes American Eagle. In your class, 11 students like Hollister, and 8 students like American Eagle. If there are 171 students in your grade, how many students like Hollister and how many like American Eagle?

Solution: To find how many students in your grade like Hollister and American Eagle, we will have to set up two proportions. In the first proportion, the number of students who like Hollister is placed over the total number of students in the class (19). Use the chart below as an organizer to help you place the information and solve for the missing information.

	Class	Grade
Like Hollister	11	X
Total	19	171

You are making a comparison between the information found in your class with a projected amount for your entire grade. We know that there are 171 students in your entire grade, but there are only 19 students in your class where you asked the question. By placing this information into the table, you have created two ratios: one for your class and one for your grade. Because the information is presented in a ratio form, you can easily find the cross products to get your solution.

$$\frac{11}{19} = \frac{x}{171}$$
$$19x = 11 \cdot 171$$
$$19x = 1,881$$
$$\frac{19x}{19} = \frac{1,881}{19}$$
$$x = 99$$

V.I.P. (Very Important Point)

As an alternate solution, you could subtract the number of students who like Hollister in the entire class and subtract that from the total number of students to find out how many students like American Eagle. This would save you some time rather than having to solve the problem by setting up two proportions.

99 people in your class like Hollister.

To find how many students like American Eagle, you can repeat the process substituting the information for American Eagle for Hollister.

	Class	Grade
Like American Eagle	8	X
Total	19	171

$$\frac{8}{19} = \frac{x}{171}$$
$$19x = 8 \cdot 171$$
$$19x = 1,368$$
$$\frac{19x}{19} = \frac{1,368}{19}$$
$$x = 72$$

72 people in your class like American Eagle.

We can check our work by adding the number of students who like Hollister (99) with the number of students who like American Eagle (72) to get the total number of students in your class (171).

SAMPLE TEST QUESTIONS (For answers, see page 178.)

Multiple-Choice (Circle the correct answer.)

1. Bart is training for a marathon. A marathon is 26.2 miles and he has run 80% of that distance. Which proportion could be used to solve for the distance he ran?

   A. $\dfrac{x}{26.2} = \dfrac{80}{100}$

   B. $\dfrac{x}{26.2} = \dfrac{100}{80}$

   C. $\dfrac{26.2}{x} = \dfrac{80}{100}$

   D. $\dfrac{80}{x} = \dfrac{100}{26.2}$

2. A rectangular garden measures 5 feet long and 8 feet wide. Your neighbor's garden is the same shape, and its length is 32 feet. How wide is your neighbor's garden?

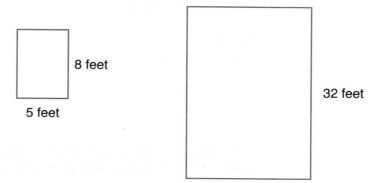

A. 5 feet

B. 20 feet

C. 26 feet

D. 30 feet

3. A cookie recipe calls for 2 cups of chocolate chips for 18 cookies. If you need to bake 54 cookies, how many cups of chocolate chips will you need?

A. 6

B. 5

C. 4

D. 3

4. Your seventh grade class reports that there is a ratio of 3 boys for every 4 girls. If there are a total 112 students in your grade, how many are boys?

A. 64

B. 49

C. 48

D. 42

5. An equilateral triangle has a perimeter of 21 inches. Xavier wants to draw an equilateral triangle with a perimeter twice as large. When he draws the new equilateral triangle, what is the length of one side?

   A. 7

   B. 14

   C. 21

   D. 42

6. A person standing 5' tall casts a shadow that is 12' long. A tree stands 15' tall. How long is its shadow?

   A. 36 feet

   B. 20 feet

   C. 15 feet

   D. 3 feet

7. A recipe that yields 10 servings of trail mix calls for 2 cups of nuts and 3 cups of raisins. If you want to make 40 servings of trail mix, how many cups of raisins do you need?

   A. 8

   B. 10

   C. 12

   D. 14

Short Constructed Response

8. Solve for x.

$$\frac{5}{8} = \frac{12.5}{x}$$

_____

LESSON 8: RATIONAL NUMBERS

Earlier in the chapter, we talked about fractions, decimals, and percents. Since you have a good knowledge of the various ways we can express numbers, let's talk about another classification of numbers called rational numbers. Rational numbers are numbers

V.I.P. (Very Important Point)

There are numbers that do not terminate or repeat when expressed as a decimal, and they are called irrational numbers. An example of an irrational number is π (pi).

that when expressed as a decimal either terminate or repeat.

It's a good idea to remember how to convert a number into decimal form to see if it terminates or repeats. If you are given a fraction, the easiest way to turn it into a decimal is to simply divide the numerator by the denominator, and you can perform that calculation either by using pencil and paper or by using a calculator.

Some rational numbers terminate, meaning that there is a remainder of zero and no more digits appear in the number. For example,

$$\frac{1}{2} = 0.5 \qquad 30\% = 0.3 \qquad \frac{2}{5} = 0.4 \qquad \sqrt{49} = 7$$

And some rational numbers, when expressed as a decimal, repeat. For example,

$$\frac{1}{3} = 0.\overline{333} \qquad \frac{1}{7} = 0.\overline{142857}$$

SAMPLE TEST QUESTIONS (For answers, see page 179.)

Multiple-Choice (Circle the correct answer.)

1. Which number is NOT between 1.0 and 1.1?

    A. 1.02
    B. 1.2
    C. 1.001
    D. 1.012

2. Which rational number is between 2.3 and 2.4?

    A. 2.33
    B. 2.41
    C. 2.28
    D. 2.035

3. While baking brownies, you can use anywhere between $\frac{3}{4}$ and 1 cup of walnuts. Which would be an acceptable amount of walnuts to use in the recipe?

   A. $\frac{5}{8}$

   B. $\frac{12}{16}$

   C. $\frac{14}{16}$

   D. $\frac{17}{16}$

4. Arrange the following rational numbers in order from LEAST to GREATEST.

   $$\sqrt{2} \quad \frac{4}{3} \quad 130\% \quad \frac{29}{20}$$

   A. $130\%, \frac{4}{3}, \sqrt{2}, \frac{29}{20}$

   B. $\frac{29}{20}, \sqrt{2}, \frac{4}{3}, 130\%$

   C. $\sqrt{2}, \frac{4}{3}, 130\%, \frac{29}{20}$

   D. $130\%, \frac{4}{3}, \frac{29}{20}, \sqrt{2}$

5. What number is between $\frac{8}{10}$ and $\frac{9}{10}$?

   A. $\frac{17}{20}$

   B. 0.79

   C. 75%

   D. $\frac{19}{20}$

6. You need a drill bit that measures $\frac{3}{8}$ inch. What is $\frac{3}{8}$ expressed as a decimal?

   A. 0.3

   B. 0.37

   C. 0.375

   D. 0.38

7. $2\frac{1}{8}$ is greater than which number?

   A. $\frac{5}{2}$

   B. 2.12

   C. 2.2

   D. $\frac{7}{3}$

Short Constructed Response

8. Write a rational number between 1.7 and 1.8. _____

Extended Constructed Response (Use the space provided to show your work.)

Draw a number line to plot and express two rational numbers that are between 0.3 and 0.4. Express one number as a decimal and the other as a fraction. Explain why these numbers are considered rational numbers.

LESSON 9: ESTIMATION AND ROUNDING

If you have ever attended the State Fair, you might wonder how many people actually attended. While it would be difficult to get an exact amount (unless you could count the tickets), you could estimate the number of people attending. While actual attendance is needed, an estimate or a projection of how many people attend such events can be helpful when ordering program books, food, drinks, and souvenirs.

When estimating, you MUST know the place value of a number's digits. Let's quickly review the place value chart.

Hundred Thousands	Ten Thousands	Thousands	Hundreds	Tens	Ones (Units)	.	Tenths	Hundredths	Thousandths

Example 1

The actual attendance at the State Fair was 58,749 people. Rounded to the nearest thousand, how many people attended the fair?

Solution: First, you must identify which digit is located in the thousands place. That number is 8. Next, look at the digit *immediately* after the thousands place (the hundreds place). If that digit is 5 or greater, add one to the thousands place and put 0s in the rest of the places. If the digit is 4 or less, leave the digit in the thousands place as it is and put zeros in the remaining places. Since the digit in the hundreds place is 7, increase the thousands place by 1 and the estimated number of people attending the fair is 59,000.

Example 2

The estimated population of Askville, New Jersey is 30,000 people. To the nearest thousand, what is the range of possible populations for the town of Askville?

Solution: To solve the problem, you must think of the "range" of values that could round to 30,000. The problem states that you need to round the number to the nearest thousand. That means if the population were 25,000 it would round up to 30,000. Also, if

V.I.P. (Very Important Point)

Don't get caught in a trap thinking that estimation problems are easy. Sometimes, they can be very tricky!

the population were 34,499 it would round down to 30,000. Therefore, since the population is being rounded to 30,000, Askville's population could be anywhere between 25,000 and 34,499 people.

Example 3

What is the estimate of 42 × 1.6?

Solution: When solving a problem like this, be careful not to enter the numbers into your calculator and solve the problem EXACTLY. This problem asks you to estimate and in a multiple-choice question, the exact answer could appear as a trick. The best way to solve a problem like this is to round each factor to its largest place value.

$$
\begin{array}{rcl}
42 & \rightarrow & 40 \\
\times 1.6 & \rightarrow & \times 2 \\
\hline
& & 80
\end{array}
$$

Example 4

What is the best estimate for the sum of 489 + 642 + 896 + 127 = ?

A. Between 0 and 1,000

B. Between 1,000 and 1,500

C. Between 1,500 and 2,000

D. Between 2,000 and 3,000

Solution: Rounding to the nearest thousand will give you two possible answers (C and D). However, if you round each number to the nearest hundred, you will determine the correct answer.

$$
\begin{array}{rcl}
489 & \rightarrow & 500 \\
642 & \rightarrow & 600 \\
896 & \rightarrow & 900 \\
+127 & \rightarrow & 100 \\
\hline
& & 2,100
\end{array}
$$

SAMPLE TEST QUESTIONS (For answers, see pages 179-180.)

Multiple-Choice (Circle the correct answer.)

1. Round 73,981 to the nearest hundred.

   A. 70,000
   B. 74,000
   C. 73,900
   D. 73,000

2. What is the best estimate for the product of 72 × 23?

   A. 700
   B. 1,050
   C. 1,400
   D. 1,656

3. At a recent game at Shea Stadium, approximately 40,000 people were in attendance. Rounded to the nearest ten thousand, what is a possible range for the actual number of people attending the game?

   A. 40,000 to 50,000
   B. 35,000 to 45,000
   C. 35,000 to 44,999
   D. 30,000 to 39,999

4. During the first week of January, you earned $74.00. The next week, you earned $42.00. Finally, you earned $55.50 during the third week. What is the best estimate for how much money you earned during those three weeks?

   A. Approximately $225.00
   B. Approximately $150.00
   C. Approximately $125.00
   D. Approximately $100.00

5. You earned $217.00 for 24 hours of work. What is the BEST way to estimate the amount of money you earned per hour?

   A. $200 ÷ 20

   B. $240 ÷ 24

   C. $210 ÷ 21

   D. $217 ÷ 21

6. Choose the best estimate for 18,982 + 56,173 + 9,641.

   A. 100,000

   B. 90,000

   C. 80,000

   D. 50,000

7. Choose the best estimate of 53 × 19.

   A. 0

   B. 500

   C. 1,000

   D. 2,000

8. Round 256,943 to the nearest thousand.

   A. 256,900

   B. 257,000

   C. 260,000

   D. 300,000

Short Constructed Response

9. Round 0.995 to the nearest hundredth. _____

Extended Constructed Response (Use the space provided to show your work.)

The estimated price of a new car is $21,500. What could be the actual range for the price of the car when it is rounded to the nearest hundred? Write an explanation using complete sentences. Use examples if necessary.

LESSON 10: PERCENT OF DISCOUNT

Do you like to save money? Hopefully you do, and are aware of the many ways to save money. Savings accounts are good, but you can also save money by not paying full price for an item at the store.

Let's imagine that the ASK Department store is advertising a sale on all televisions. Any television in their showroom is on sale for 25% off the regular price. So, if a new television set normally sells for $900, you can purchase that same television for 25% off the regular price. How would you calculate the sale price of the television set?

Example 1

A $900 television is on sale for 25% off the regular price. What is the sale price?

Solution: To calculate the sale price, set up a proportion similar to the proportions we solved in a previous section.

$$\frac{is}{of} = \frac{\%}{100}$$

$$\frac{x}{900} = \frac{75}{100}$$

Since you are paying 75% of the regular price of the television set, this proportion will allow you to determine your answer.

$$100x = 900 \cdot 75$$
$$100x = 67,500$$
$$\frac{100x}{100} = \frac{67,500}{100}$$
$$x = 675$$

The television's sale price is $675.

To calculate a sale price, you can think of it in terms of this formula:

Regular Price × Percent of Discount = Amount of Discount

Regular Price – Amount of Discount = Sale Price

V.I.P. (Very Important Point)

When solving these problems, you must identify the four key pieces of information.

Regular Price = the normal selling price of the item.

Percent of Discount = a percentage that, when multiplied by the regular price, will equal the amount of discount.

Amount of Discount = the percentage found by multiplying the regular price by the percent of discount. This amount is subtracted from the regular price to determine the sale price.

Sale Price = the final selling price of an item, which is calculated by subtracting the amount of discount from the regular price.

Now, what if you knew the sale price and the regular price and you wanted to determine the percent of discount? As long as you know those two pieces of information, you can calculate the percent of discount easily.

Example 2

A portable DVD player is on sale for $150. It normally sells for $250. What is the percent of discount?

Solution: Looking back at our formula, you need to find the amount of discount. This is determined by subtracting the sale price from the regular price. (In other words, what is the difference between the sale price and the regular price?)

Regular Price: $250

Sale Price: $150

Amount of Discount: $100   ($250 – $150 = $100)

Remember that the amount of discount represents the percentage of the regular price that was discounted. So, to calculate the percent of discount, set up a proportion as we did in an earlier section.

$$\frac{\text{Amount of Discount}}{\text{Regular Price}} = \frac{x}{100}$$

$$\frac{\$100}{\$250} = \frac{x}{100}$$
$$\$250x = \$100 \cdot 100$$
$$\frac{\$250x}{\$250} = \frac{\$10,000}{\$250}$$
$$x = 40$$

The portable DVD player is 40% off its regular price.

SAMPLE TEST QUESTIONS (For answers, see pages 180–181.)

Multiple-Choice (Circle the correct answer.)

1. A pair of sneakers normally sells for $60.00. It is on sale for $39.00. What is the percent of discount?

   A. 21%

   B. 30%

   C. 35%

   D. 40%

2. Jamal is shopping for a CD. He wants to buy a CD that normally sells for $15.00 but is on sale for 20% off. If he buys the CD, how much will he pay for it?

   A. $14.80

   B. $12.00

   C. $9.00

   D. $3.00

3. Cathy is buying new living room furniture. She wants to purchase a set that costs $3,200, but she only has $3,000 in her budget. What percent must the furniture go on sale for in order for Cathy to purchase the furniture?

   A. 4%

   B. 5%

   C. 6%

   D. 7%

4. On a shopping trip to a local department store, Josephine purchases a shirt for $20.00, a pair of jeans for $35.00, and a pair of sneakers for $50.00. At the register, she redeems a coupon for 30% off her entire bill. With the discount, how much does she pay in total?

   A. $73.50

   B. $75.00

   C. $104.70

   D. $105.00

5. Jake and Luke go to the store to buy camping equipment. Their total cost for a tent, sleeping bags, and supplies is $270.00, which reflected a 20% discount. What was the original price of the camping equipment?

   A. $54.00

   B. $250.00

   C. $290.00

   D. $337.50

6. What is the percent of discount on a television that originally sold for $700 and is now on sale for $560?

   A. 20%

   B. 80%

   C. 140%

   D. 560%

7. A media player originally sells for $200. It is on sale for 30% off. What is the sale price?

   A. $30

   B. $60

   C. $140

   D. 170

Short Constructed Response

8. Find the percent of discount when the regular price for a DVD recorder is $250 and its sale price is $175. _____

Extended Constructed Response (Use the space provided to show your work.)

Jameka bought a $60.00 necklace on sale for 40% off its regular price. Caitlyn bought an $80.00 necklace for 55% off its regular price. Who paid the least amount for their necklace? Show your work and explain your reasoning.

LESSON 11: SQUARE NUMBERS AND CUBIC NUMBERS

A square number is the product of a number multiplied by itself. If n represents any positive integer, then n^2 would represent a square number. Also, a square number is a number in which the amount of dots representing that number could be arranged in a square. For example, 9 is a square number because I could take 9 dots and arrange them in such a way to form a 3×3 square.

●　　●　　●

●　　●　　●

●　　●　　●

V.I.P. (Very Important Point)

When you take the square root of a square number, the result is an integer.

A cubic number is the product of a number multiplied by itself three times. For example, if n were any positive integer, then n^3 would represent a cubic number.

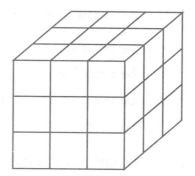

Here is a list of some square and cubic numbers. Familiarize yourself with these numbers so you can easily recognize them.

The First 10 Square Numbers: 1, 4, 9, 16, 25, 36, 49, 64, 81, and 100

The First 10 Cubic Numbers: 1, 8, 27, 64, 125, 216, 343, 512, 729, and 1,000

SAMPLE TEST QUESTIONS (For answers, see page 181.)

Multiple-Choice (Circle the correct answer.)

1. Which square number is greater than 50?

   A. 5^2

   B. 6^2

   C. 7^2

   D. 8^2

2. Which cubic number is equivalent to 343?

   A. 8^3

   B. 7^3

   C. 6^3

   D. 5^3

3. If $n < 0$, then n^3 would

    A. be positive.

    B. be negative.

    C. be positive or negative.

    D. equal zero.

4. Which number is a cubic number greater than 70?

    A. 81

    B. 125

    C. 144

    D. 200

Short Constructed Response

5. What is the value of x when $x^3 = 1,331$?   _____

Extended Constructed Response (Use the space provided to show your work.)

Identify any square numbers and cubic numbers that are greater than 0 and less than 100. Identify any number(s) that is a square number and a cubic number.

LESSON 12: DIVISIBILITY

You are going to throw a party for 28 people. While you are making arrangements for the music and the party favors, you will need tables for everyone to sit and enjoy something to eat. How are you going to get enough tables for everyone? How many people will you sit at each table? Since you don't have enough tables at your house, you need to contact a local party rental store and rent tables for the party.

The local party store tells you that you can rent tables that seat 4 people or 8 people. Since there are 28 people coming to your party, which tables will you choose? You could rent 7 tables that seat 4 people per table because $7 \times 4 = 28$ and that is the *exact* number of people attending your party. It wouldn't make sense to rent the tables that seat 8 because you would have some empty seats. For example, if you rent 3 tables that seat 8 people that would only seat 24 people. But, if you rented 4 tables that seat 8 people, that would seat 32 people and that would be too much. Since 28 is evenly divisible by 4, it is best for you to rent 7 tables that seat 4 people.

You were able to solve this problem because you knew that 28 was divisible by 4. When we say that a number is divisible by 4, we mean that a given number divided by 4 will yield a whole number as its answer. In other words, there is no remainder. Let's say if 30 people were coming to your party, it wouldn't make sense to have everyone sitting at tables of four because you would have 7 full tables with an additional table where only two people would sit. That wouldn't be good planning and wouldn't be very nice for your guests. If 30 people were coming instead of 28, what size tables would you want at your party?

To solve divisibility problems, you can learn these very easy rules as a mental math trick to help you solve them.

A Number Is Divisible by:	Explanation	Example
2	A number is divisible by 2 if the number is even. The units digit is an even number.	24 is divisible by 2 because $24 \div 2 = 12$
3	A number is divisible by 3 if the sum of the number's digits is divisible by 3.	183 is divisible by 3 because the sum of its digits is 12 $(1 + 8 + 3 = 12)$ and 12 is divisible by 3.

A Number Is Divisible by:	Explanation	Example
4	A number is divisible by 4 if the number formed by the last two digits is divisible by 4.	548 is divisible by 4 because 48 is divisible by 4.
5	A number is divisible by 5 if its units digit is 5 or 0.	$370 \div 5 = 74$ $825 \div 5 = 165$
6	A number is divisible by 6 if it is divisible by 2 and 3.	162 is divisible by 6 because its units digit is 2 and the sum of its digits is divisible by 3 $(162 \div 6 = 27)$.
8	A number is divisible by 8 if the number formed by the last three digits is divisible by 8.	1,440 is divisible by 8 because $440 \div 8 = 55$ $(1,440 \div 8 = 180)$.
9	A number is divisible by 9 if the sum of its digits is divisible by 9.	297 is divisible by 9 because $2+9+7=18$ and 18 is divisible by 9 $(297 \div 9 = 33)$.
10	A number is divisible by 10 if the units digit is 0.	$780 \div 10 = 78$

Let's try some problems and use the rules to help us solve the problems quickly and accurately.

Example 1

Is 484 divisible by 6?

V.I.P. (Very Important Point)

When you examine the above list of divisibility rules, you will notice that a rule for 7 has been omitted. There is a rule but it involves a lot more steps than those described. Since the NJ ASK7 test is calculator-ready, you can simply enter the number into the calculator and divide by 7. If the answer revealed is a whole number, then the number is divisible by 7. It's also good to remember that if you happen to forget the divisibility rules, you can always use your calculator to help you solve the problem.

Solution: For a number to be divisible by 6, it must be divisible by 2 and 3. While 484 is an even number and all even numbers are divisible by 2, it is not divisible by 3. 484 is not divisible by 3 because the sum of its digits is 16, and 16 is not divisible by 3. Therefore, 484 is not divisible by 6.

Example 2

Determine whether 2,148 is divisible by 2, 3, 4, 5, 6, 8, 9, or 10.

Solution: Using the divisibility rules, 2,148 is divisible by

- 2 because it is even
- 3 because the sum of the digits (2 + 1 + 4 + 8 = 15) is divisible by 3
- 4 because the number formed from the last two digits (48) is divisible by 4
- 6 because the number is divisible by 2 and 3

However, 2,148 is NOT divisible by

- 5 because the units digit is not a 5 or 0
- 8 because the number formed by the last three digits (148) is not divisible by 8
- 9 because the sum of the digits (15) is not divisible by 9
- 10 because the units digit is not a 0

Example 3

Write a number that is divisible by 3, 4, and 5.

Solution: We can use some logical reasoning to solve this problem rather than simply guessing and checking. If a number is divisible by 3, 4, and 5, it means that the number must have factors of 3, 4, and 5. Since a factor is a number used in multiplication, we can multiply 3 × 4 × 5 to get an answer. That answer is 60. Of course, more numbers satisfy this problem, but you only have to give one number as an acceptable answer.

SAMPLE TEST QUESTIONS (For answers, see page 182.)

Multiple-Choice (Circle the correct answer.)

1. 3,894 is divisible by

   A. 2

   B. 3

   C. 6

   D. All of the above

2. A local pizzeria has tables that seat 4 or 6 people. A party of 38 people comes into the restaurant. How can you sit these 38 people without having any empty seats at any tables?

    A. 2 tables of 4 and 6 tables of 6

    B. 3 tables of 4 and 5 tables of 6

    C. 4 tables of 5 and 3 tables of 6

    D. 6 tables of 4 and 2 tables of 6

3. Which number is divisible by 2, 4, and 9?

    A. 144

    B. 90

    C. 54

    D. 44

4. Which number is NOT divisible by 7?

    A. 84

    B. 154

    C. 197

    D. 252

5. The seventh grade is going on a class trip to Six Flags Great Adventure. There are a total of 176 students and teachers going on the trip. Each bus can hold 44 people. How many buses will you need for the trip?

    A. 2

    B. 3

    C. 4

    D. 5

6. Your physical education teachers are going to play a game with the seventh grade class, and each team needs to consist of 6 people. Which class size will allow your teachers to create even teams?

   A. 150
   B. 153
   C. 158
   D. 159

7. Your seventh grade class is going on a trip to the New Jersey Performing Arts Center. Including teachers, 32 people will go on the trip. The bus has seats that hold 2 people (double seats) and some that hold 3 people (triple seats). Which seating arrangement is possible on the bus?

   A. 8 double seats and 5 triple seats
   B. 7 double seats and 6 triple seats
   C. 6 double seats and 7 triple seats
   D. 5 double seats and 8 triple seats

Short Constructed Response

8. Write a two-digit number greater than 20 but less than 40 that is divisible by 3 but not by 9. _____

Extended Constructed Response (Use the space provided to show your work.)

Your teacher needs to divide your class of 23 students into even groups. Explain how your teacher can do this. Use complete sentences and/or a diagram to support your explanation.

LESSON 13: PRIME AND COMPOSITE NUMBERS / PRIME FACTORIZATION

Within our number system, positive integers greater than 1 are considered either prime or composite. A prime number is a number whose factors are only 1 and itself. A composite number is a number whose factors are 1, itself, and at least one other factor. Remember, a factor is a number you multiply.

Can you name some prime numbers between 1 and 20? Think of a series of numbers whose factors are only 1 and itself.

Perhaps you can conclude that 2, 3, 5, 7, 11, 13, 17, and 19 are prime numbers. Do you notice that 2 is the only even prime number? That's important to know because any other even number, or any multiple of 2, has to be a composite number. Maybe, after examining this list, you also see that 9 is not a prime number because its factors are 1, 9, and 3 (because $3 \times 3 = 9$).

Any composite number can be expressed as a product of its prime factors. This is called prime factorization. For example, what is the prime factorization of 24? Since 24 is an even number, I know that it must have a factor of 2. Therefore,

24	
2×12	24 is even so $24 = 2 \times 12$
2×6	12 is even so $12 = 2 \times 6$
2×3	6 is even so $6 = 2 \times 3$

Personally, I think it is much easier to perform prime factorization when you start with the number 2. By that I mean, if the number is even, it MUST have a factor of 2. So, to determine its prime factors, keep factoring out a 2 until you can't do it any more. In the example above, $24 = 2 \times 12$. Since 12 is even, $12 = 2 \times 6$. Then, since 6 is even, $6 = 2 \times 3$. I know that 2 and 3 are prime numbers so the prime factorization of $24 = 2 \times 2 \times 2 \times 3$. It is also good practice to express the prime factorization of any number using exponents. For example, $24 = 2^3 \times 3$.

Let's practice finding the prime factorization of some other numbers.

Example 1

Find the prime factorization of 48.

48

2×24	48 is even so $48 = 2 \times 24$
2×12	24 is even so $24 = 2 \times 12$
2×6	12 is even so $12 = 2 \times 6$
2×3	6 is even so $6 = 2 \times 3$

The prime factorization of $48 = 2 \times 2 \times 2 \times 2 \times 3$. You can also express the prime factorization of 48 as $2^4 \times 3$. *Hint*: If it helps, circle the prime numbers in the prime factorization.

Example 2

Find the prime factorization of 189.

Solution: Since I know that 189 is not even, it can't have a prime factor of 2. In the previous section, we learned that any number whose digits sum to a multiple of 3 is divisible by 3. Since $1 + 8 + 9 = 18$ and 18 is a multiple of 3, 189 is divisible by 3 and I can then perform its prime factorization.

189

3×63	189 is divisible by 3 so $189 = 3 \times 63$
3×21	63 is divisible by 3 so $63 = 3 \times 21$
3×7	21 is divisible by 3 so $21 = 3 \times 7$

The prime factorization of 189 is $3 \times 3 \times 3 \times 7$ or $3^3 \times 7$.

V.I.P. (Very Important Point)

The number 1 is considered neither prime nor composite. It is considered a special case because it doesn't satisfy either definition. Remember, the only factor of 1 is 1. So, since 1 is itself, it can't be prime or composite. Sometimes, questions will try to trick you into thinking that 1 is a prime number, but now you know better and will not fall into that trap!

Example 3

Of 21, 27, 40, and 70, which number has the most number of DIFFERENT prime factors?

Solution: This is a great multiple-choice problem that will require you to find the prime factorization of each number. Remember, the key word here is *different*. It's bolded to make you notice that it is important to the problem. So, let's find the prime factorization of each number.

$21 = 3 \times 7$	There are 2 different prime factors: 3 and 7.
$27 = 3 \times 3 \times 3$	There is only 1 prime factor here and that is 3.
$40 = 2 \times 2 \times 2 \times 5$	There are 2 different prime factors: 2 and 5.
$70 = 2 \times 5 \times 7$	There are 3 different prime factors: 2, 5, and 7.

Since 70 has three different prime factors and has the most different prime factors in the group, that is the correct answer.

SAMPLE TEST QUESTIONS (For answers, see pages 182-183.)

Multiple-Choice (Circle the correct answer.)

1. Which number has the greatest number of DIFFERENT prime factors?

    A. 10

    B. 20

    C. 30

    D. 40

2. Which number has the LEAST number of different prime factors?

    A. 38

    B. 50

    C. 82

    D. 121

3. The prime factorization of $48 = 2^? \times 3$ is

    A. 5

    B. 4

    C. 3

    D. 2

4. What is the correct prime factorization of 52?

    A. 2×26

    B. $2 \times 2 \times 13$

    C. 1×52

    D. 52 is a prime number

5. Which number is the largest prime number less than 80?

    A. 71

    B. 73

    C. 77

    D. 79

6. What is the only prime number that is greater than 25 but less than 30?

    A. 26

    B. 27

    C. 28

    D. 29

7. Which number is NOT a prime number?

    A. 121

    B. 119

    C. 113

    D. 103

8. Which prime number is greater than 50 but less than 60?

    A. 51

    B. 53

    C. 55

    D. 58

9. Which number has the most number of DIFFERENT prime factors?

    A. 24

    B. 35

    C. 45

    D. 70

Short Constructed Response

10. Write a composite number that has a factor of 2, 3, and 5. _____

Extended Constructed Response

Goldbach's Conjecture states that any number greater than 2 can be expressed as the sum of two prime numbers. Given the following numbers, use Goldbach's Conjecture to show that these numbers are the sum of two prime numbers.

18 = _____

28 = _____

40 = _____

Chapter 2

GEOMETRY AND MEASUREMENT

LESSON 1: TYPES OF POLYGONS

You can see many different shapes simply by looking around your classroom or your home. Shapes come in different sizes and forms, but they share one common bond. They are all closed figures made up of line segments. In mathematics, we define a polygon as a closed figure made up of three or more line segments. A regular polygon is a polygon with congruent line segments. Take a moment to look around the room and see how many different shapes you can identify.

Examine and study the following list of polygons. You should know these figures and memorize their characteristics.

Polygon	Characteristics	Total Degrees	Measure of Each Interior Angle (Based on a Regular Polygon = Total Degrees ÷ Number of Sides)
Triangle	3 Sides Classified by sides (Scalene—zero equal sides; Isosceles—two equal sides; Equilateral—three equal sides). Classified by angles (Acute—all angles less than 90°; Right—one angle measures 90°; Obtuse—one angle measures more than 90°)	180°	60°

Polygon	Characteristics	Total Degrees	Measure of Each Interior Angle
Quadrilateral	4 sides Types: Parallelogram, Rectangle, Rhombus, Square, Trapezoid. A trapezoid is the only quadrilateral that has only one pair of parallel sides.	360°	90°
Pentagon	5 sides Think of the government building in Washington, DC	540°	108°
Hexagon	6 sides	720°	120°
Octagon	8 sides This shape is used as a STOP sign.	1,080°	135°
Nonagon	9 sides	1,260°	140°
Decagon	10 sides	1,440°	144°
Dodecagon	12 sides	1,800°	≈157°

***

V.I.P. (Very Important Point)

To calculate the total number of degrees for any-sided polygon find the number of sides of the polygon and subtract 2, then multiply by 180.

Total Degrees of Polygon =
180 (Number of Sides − 2)

***

When answering questions on the NJ ASK7 pertaining to polygons, you should be familiar with what they look like, how many sides they have, and how many total degrees are in each polygon. We stated earlier that a triangle has 180° and any quadrilateral has 360°. What is the difference between the total number of degrees in a quadrilateral and the total number of degrees in a triangle? The answer is 180°. It is good to know that there is a pattern between the number of sides in a polygon and its total number of degrees.

Example

Find the total number of degrees in a heptagon.

Solution: To calculate the total number of degrees, subtract 2 from the number of sides and multiply by 180°. Since a heptagon is a 7-sided polygon,

$$180(7 - 2)$$
$$= 180 \cdot 5$$
$$= 900$$

There are 900° in a heptagon.

SAMPLE TEST QUESTIONS (For answers, see pages 183–184.)

Multiple-Choice (Circle the correct answer.)

1. Marcus is playing baseball at a local park. He and his friends refer to the playing field as a diamond but it is classified as which kind of polygon?

    A. Triangle
    B. Quadrilateral
    C. Pentagon
    D. Hexagon

2. This quadrilateral has two pairs of parallel sides and two sets of congruent angles that measure 135° and 45°. What type of polygon is it?

    A. Square
    B. Rectangle
    C. Parallelogram
    D. Trapezoid

3. In $\square ABCD$, $m\angle A = 115°$, what is $m\angle C$?

    A. 115°
    B. 65°
    C. 45°
    D. 25°

4. I am a quadrilateral that has only one pair of parallel sides. What quadrilateral am I?

   A. Rectangle

   B. Rhombus

   C. Square

   D. Trapezoid

5. How many total degrees are there in a decagon?

   A. 900°

   B. 1,080°

   C. 1,260°

   D. 1,440°

6. What do you call a polygon with 12 sides?

   A. Dodecagon

   B. Dozecagon

   C. Decagon

   D. Septagon

Short Constructed Response

7. A pentagon has three angles measuring 140°, 80°, and 110°, respectively. If the other two angles are congruent, what is the measure of each angle?   _____

Extended Constructed Response (Use the space provided to show your work.)

Using the triangles found on your Mathematics Reference Sheet, how many triangles will you need to make the following figures? Trace them on your paper and write the number of triangles used to construct each figure.

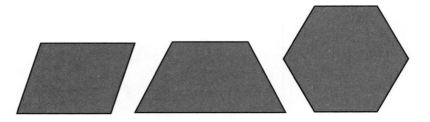

LESSON 2: LINES AND ANGLES

Have you ever heard the expression "The shortest distance between two points is a straight line"? When studying geometry it is absolutely true. Two points through which a line passes continuing in opposite directions always define any straight line.

If you have two points, called *A* and *B*, they can define a straight line.

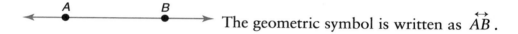 The geometric symbol is written as $\overleftrightarrow{AB}$.

A line segment is only part of a line.

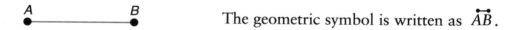 The geometric symbol is written as $\overline{AB}$.

Sometimes you can have a line that extends in only one direction. This is called a ray.

A ray has an endpoint and from that endpoint, the line extends infinitely in one direction. It never stops going but it only goes in one direction.

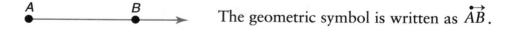 The geometric symbol is written as $\overrightarrow{AB}$.

Whenever you have two rays drawn together so that they share a common endpoint or vertex, it forms an angle.

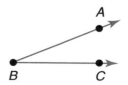

V.I.P. (Very Important Point)

Here are some hints and helpful study aids to remember the kinds of angles.

Acute—It's a tiny angle. Think of it as a small angle that is very "cute."

Right—This angle looks like the letter "L" or a backwards "L." There is usually a box at the vertex to indicate that it is a right angle.

Obtuse—The angle is a large angle. The angle looks as if it is open wide.

Straight—This angle is a "straight" line.

The geometric symbol is written as ∠ABC.

An angle is named by using three points. The middle letter is always the vertex. In the above example, the name of that angle is ∠ABC.

There are four types of angles. The angle measurements can be determined by using a protractor, but sometimes you can use algebraic reasoning to determine their measurement.

The four types of angles are

Acute—any angle measure which is greater than 0° and less than 90°

Right—any angle measure that equals 90°

Obtuse—any angle measure that is greater than 90° and less than 180°

Straight—any angle measure that equals 180°

Example 1

Identify the line shown.

Solution:

$$\overleftrightarrow{MN}$$

Example 2

Identify the following angle by name and its classification.

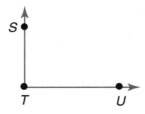

Solution: The name of the angle is ∠STU, and it is a right angle.

V.I.P. (Very Important Point)

Complementary Angles = Two angles whose sum is 90°

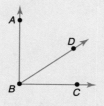

$\angle ABC = 90$

$\angle ABD = 60$

$\angle DBC = 30$

Since $\angle ABC + \angle DBC$ equal 90°, $\angle ABD$ and $\angle DBC$ are complementary.

Supplementary Angles = Two angles whose sum is 180°

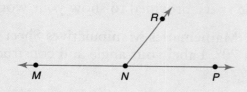

$\angle MNP = 180$

$\angle MNR = 120$

$\angle RNP = 60$

Since $\angle MNR + \angle RNP = 180°$, $\angle MNR$ and $\angle RNP$ are supplementary.

An easy way to remember this is to think of them alphabetically and numerically.

Complementary comes before supplementary. 90° come before 180°.

Complementary angles = 90°
Supplementary angles = 180°

SAMPLE TEST QUESTIONS
(For answers, see page 184.)

Multiple-Choice (Circle the correct answer.)

1. Karen is making a picture frame using wood she has in her garage. If she needs to put the frame around a rectangular picture, the wood must be cut on a 45° angle. What kind of angle is a 45° angle?

   A. Acute
   B. Right
   C. Straight
   D. Obtuse

2. When two rays share a common vertex and the ends are in opposite directions, it forms what kind of angle?

   A. Acute
   B. Right
   C. Straight
   D. Obtuse

3. What is the correct way to name the following diagram?

   A. $\overleftrightarrow{SQ}$
   B. $\overline{SQ}$
   C. $\overgroup{SQ}$
   D. $\angle SQ$

4. The m∠*JKL* = 120°. Its supplement is 2*x* + 10. What is the value of *x*?

   A. 5

   B. 10

   C. 25

   D. 60

5. The m∠*XYZ* = 5*n*+ 50. ∠*XYZ* is a right angle. What is the value of *n*?

   A. 90

   B. 10

   C. 8

   D. 5

Extended Constructed Response (Use the space provided to show your work.)

Use your protractor from your Mathematics Manipulatives Sheet to construct ∠*ASK* that measures 70°. Label your angle and construct its complement.

LESSON 3: SIMILAR AND CONGRUENT FIGURES

These isosceles trapezoids are congruent because they are the same shape; they have the same angle measures and are the same size. We can say that trapezoid *ABCD* ≅ trapezoid *EFGH*.

Rectangle *STUV* and rectangle *WXYZ* are not the exact size and shape. However, they do have the same angle measures. When two figures are the same shape and have the same angle measures but are different sizes, they are similar. We can say that rectangle *STUV* ~ rectangle *WXYZ*.

Similar figures are in proportion with each other. This means that we can create a proportion with the ratios representing the relationship between two dimensions of the figure and their cross products are equal. Let's look at an example to clarify this.

Example 1

Determine if the following two figures are similar by using proportions.

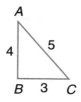

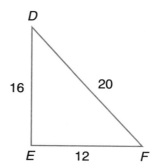

Solution: If $\triangle ABC \sim \triangle DEF$, then $\dfrac{AB}{DE} = \dfrac{BC}{EF}$. If the cross products

are equal, then the two figures are similar.

So, $\dfrac{4}{16} = \dfrac{3}{12}$ and to prove by cross products, $4 \cdot 12 = 48$ and

$3 \cdot 16 = 48$. Since the cross products are equal, we can conclude that the two figures are similar.

Example 2

The following figures are similar. Find the value of x.

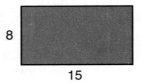

8

15

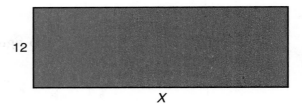

12

X

Solution: To find the value of X, set up a proportion and solve.

$$\frac{8}{12} = \frac{15}{X}$$
$$8X = 15 \cdot 12$$
$$8X = 180$$
$$X = 22.5$$

SAMPLE TEST QUESTIONS (For answers, see pages 184-185.)

Multiple-Choice (Circle the correct answer.)

1. △*EFG* is an equilateral triangle with a side measurement of 15 mm. △*HIJ* is also an equilateral triangle that is three times as large as △*EFG*. What is the measure of each side of △*HIJ*?

    A. 5 mm

    B. 15 mm

    C. 30 mm

    D. 45 mm

2. A 4 in. × 6 in. picture frame is NOT similar to which picture frame?

    A. 2 in. × 3 in.

    B. 6 in. × 8 in.

    C. 8 in. × 12 in.

    D. 16 in. × 24 in.

3. □*EFGH* ~ □*LMNO* If *EF* measures 8 in. and *LM* measures 16 in., how many times larger is the perimeter of □*LMNO* than the perimeter of □*EFGH*?

    A. 16

    B. 8

    C. 4

    D. 2

4. Refer to question 3. How many times larger is the area of □*LMNO* compared with the area of □*EFGH*?

    A. 16

    B. 8

    C. 4

    D. 2

5. Two figures are congruent. Which statement is true?

   A. Congruent figures are the same size.

   B. Congruent figures have the same angle measures.

   C. Congruent figures have the same area.

   D. All of the above

6. Two figures are similar. They have

   A. the same perimeter.

   B. the same area.

   C. the same angle measurements.

   D. the same side measurements.

Short Constructed Response

7. A square's perimeter measures 64 in. If a similar square's dimensions are $\frac{1}{2}$ as large, what is the measure of each side? _____

LESSON 4: CIRCLES

A circle is perfect. It's round. It has no edges. Every point on the circle is an equal distance from the center point. And most importantly, it is NOT a polygon. Remember that a polygon is a closed figure made up of line segments, but a circle doesn't have any line segments.

A circle has many parts that you need to know, and it all starts with a center point.

Parts of a Circle	Definition
Center point	A point located in the middle of a circle, which is an equal distance from every point located on the outside of the circle
Radius	A line segment that connects the center point to one point on the outside of the circle; it is $\frac{1}{2}$ of the diameter
Diameter	A line segment that connects two points on the circle and passes through the center point; a diameter is always twice the length of the radius
Chord	A line segment that connects two points on the circle; it does not have to pass through the center point. Therefore, a diameter is also a chord.

$\overline{CD}$ = Diameter

$\overline{PR}$ = Radius

$\overline{AB}$ = Chord

P = Center Point

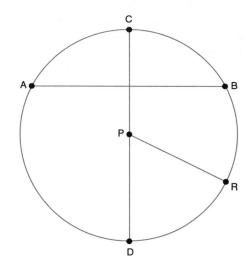

V.I.P. (Very Important Point)

Even though you have a reference sheet to help you with the formulas for NJ ASK7, here are a couple of friendly rhymes to help you remember how to calculate the area and circumference of a circle.

Area = Fuzzy Wuzzy was a bear.
Area of a circle is πr^2.

Circumference = Twinkle, Twinkle, little star, circumference of a circle is $2\pi r$.

Since the diameter is also twice the radius, you could change the circumference formula to $C = \pi d$.

When understanding the many components of a circle, it is also important for you to know how to calculate the area and circumference of a circle. Circumference is another way of describing the distance around a figure. You could think of it as being similar to perimeter, but we can't call it perimeter because that relates to polygons and circles are not polygons.

Example 1

What is the circumference and area of the circle?

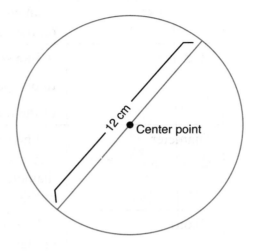

Solution:

Circumference = $2\pi r$ or πd. So, $C = 12\pi$. Since $\pi \approx 3.14$, the circumference = $12 \cdot 3.14$ = 37.68 inches.

$$Area = \pi r^2$$

$$Area = 3.14 \cdot 6^2$$

$$Area = 3.14 \cdot 36 = 113.04 \text{ in.}^2$$

Example 2

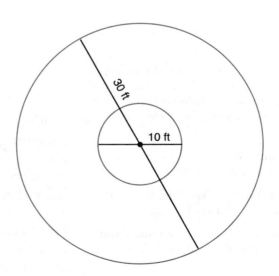

A circular walkway is constructed at a local museum. The inner circle, whose space is used as a garden, has a diameter of 10 feet. The outer circle, whose space is the actual walkway, has a diameter of 30 feet. What is the actual area of the walkway?

Solution: To calculate the area of the walkway, we will need to find the area of the larger circle and then subtract the area of the smaller circle. In order to do this, we must treat each circle separately.

$$\text{Outer circle's area} = \pi r^2$$
$$= 3.14 \cdot 15^2$$
$$= 3.14 \cdot 225$$
$$= 706.5 \text{ ft}^2$$

$$\text{Inner circle's area} = \pi r^2$$
$$= 3.14 \cdot 5^2$$
$$= 3.14 \cdot 25$$
$$= 78.5 \text{ ft}^2$$

$$\text{Outer circle's area} - \text{Inner circle's area} = \text{Area of walkway}$$
$$706.5 \text{ ft}^2 - 78.5 \text{ ft}^2 = 628 \text{ ft}^2$$

The area of the walkway is 628 ft^2.

SAMPLE TEST QUESTIONS (For answers, see page 185.)

Multiple-Choice (Circle the correct answer.)

1. A wagon wheel has a diameter of 3 feet. If the wheel rotates 5 times, how far has the wheel traveled?

   A. Approximately 9 feet

   B. Approximately 15 feet

   C. Approximately 47 feet

   D. Approximately 94 feet

2. An interior designer at a local mall has constructed a circular water fountain in the courtyard. You are assigned to tile the bottom of the fountain. If the diameter measures 40 feet, what is the area of the bottom of the fountain?

   A. 1,256 ft²

   B. 125.6 ft²

   C. 12.56 ft²

   D. 1.256 ft²

3.

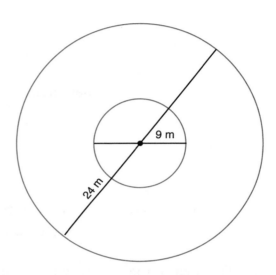

   A circular driveway features an inner circle where a statue is displayed. What is the approximate area of the driveway?

   A. 254 m²

   B. 389 m²

   C. 763 m²

   D. 1,554 m²

4. A circle's area is approximately 380 in.². What is the circle's radius?

   A. Between 5 and 10 in.

   B. Between 10 and 15 in.

   C. Between 15 and 20 in.

   D. Between 20 and 25 in.

5. If a circle's radius is tripled, what happens to the circle's area?

   A. The area is cut by $\frac{1}{3}$.

   B. The area is tripled.

   C. The area is multiplied by 9.

   D. The area is multiplied by 27.

Short Constructed Response

6. What is the radius of a circle that has an area of 153.86 cm^2? _____

LESSON 5: TRANSFORMATIONS: TRANSLATIONS, ROTATIONS, AND REFLECTIONS

Have you ever had to rearrange a room in your house? Maybe you just started moving some furniture or perhaps decided to turn some things around in the room. How did the room look when you were finished? I'm willing to guess that the room looked a bit different but that the furniture you moved was still the same shape and size but was just moved to a different location.

In geometry, we move objects and shapes around by performing transformations. A transformation is a general category for three types of moves. Those moves are called translations, rotations, and reflections. In this lesson, we will explore these three kinds of transformations and see how they relate to the questions posed on the NJ ASK7.

TRANSLATIONS

When you translate a figure, you need to move it or slide it in a particular direction. The shape always stays the same but is moved to a different location either on the coordinate plane or across some type of line.

V.I.P. (Very Important Point)

Here are some key words to help you remember what each transformation does to a figure.

Transformation	Key Word
Translation	Slide
Rotation	Turn
Reflection	Flip

Example 1

△*ABC* is located on a coordinate plane with the following ordered pairs: *A*(–2, 1), *B*(–2, –2), and *C*(1, –2). The triangle is translated three units up and two units to the right. What are the new dimensions of the triangle?

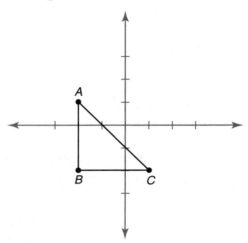

V.I.P. (Very Important Point)
In the above solution, you will notice that the new points are labeled *A′*, *B′*, and *C′*. This is a way to indicate the new ordered pair after the transformation. *A′* is read as "*A* prime."

Solution: To solve the problem, take each point and move it based on the translation described in the problem. The translation states that each point will be moved three units up and two units to the right. Therefore, the new points would be *A′*(0, 4), *B′* (0, 1), and *C′* (3, 1).

ROTATION

When you rotate a figure, it is as if you are taking one vertex of the figure and turning it. These turns can be described as either fractions or by degrees. For example, a quarter turn would be described as $\frac{1}{4} = 90°$, a half turn would be described as $\frac{1}{2} = 180°$, and a three-quarter turn would be described as $\frac{3}{4} = 270°$.

Example 2

Examine this shape.

What would the shape look like if it was rotated 90°?

Solution:

REFLECTION

When you look in the mirror, you see your reflection. A reflection is an image from a different point of view. Mirrors are great tools for helping us see reflections and can be very useful in your math classes to examine reflecting shapes and objects. When reflecting a figure you must have a line, which is used as the reflection spot. When answering problems about reflecting figures on the coordinate plane, the x- or y-axis will be used to reflect that figure. Keep in mind that when you reflect the figure, however many spaces are between the original figure and the axis must be the same once the figure is reflected. Let's take a look at an example.

Example 3

$\triangle ABC$ is drawn on the coordinate plane. Draw its reflected image when it is reflected over the y-axis.

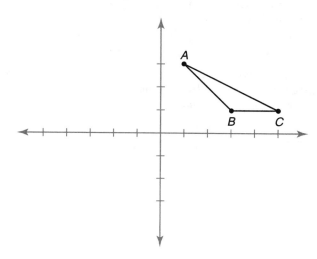

Solution: Imagine a mirror sitting right on the *y*-axis. Where would the triangle be placed? Look at the reflected image and notice how the exact spacing between the vertices and the axes are the same.

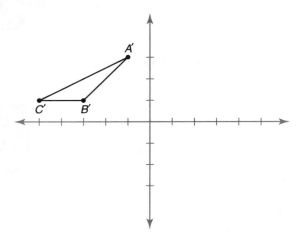

Also, look at the ordered pairs and notice what is different between the vertices of the original image and the reflected image.

Original Image	Reflected Image
A (1, 3)	*A'* (−1, 3)
B (3, 1)	*B'* (−3, 1)
C (5, 1)	*C'* (−5, 1)

V.I.P. (Very Important Point)

When reflecting images on the coordinate plane, remember this very important point.

If the image is being reflected over the *x*-axis, the reflected image is determined by keeping the original *x*-coordinate and taking the opposite *y*-coordinate.

If the image is being reflected over the *y*-axis, the reflected image is determined by keeping the original *y*-coordinate and taking the opposite *x*-coordinate.

The only coordinate that changed when reflecting the image over the *y*-axis was the *x*-coordinate.

SAMPLE TEST QUESTIONS (For answers, see page 185.)

Multiple-Choice (Circle the correct answer.)

1.

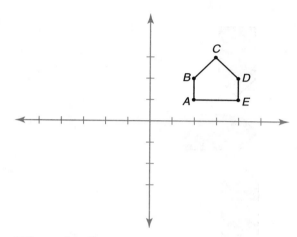

When the figure is translated 5 units down and 4 units left, what is the translated image of point C?

A. (0, −2)

B. (−1, −2)

C. (−2, −1)

D. (0, −3)

2. Refer to question 1. If the figure was reflected over the y-axis, what would be the reflected image of point E?

A. (−4, 2)

B. (−3, 3)

C. (−4, 2)

D. (−4, 1)

3.

When the figure is rotated 90°, what is its rotated image?

A.

B.

C.

D.

4. Martin is 5 blocks south and 3 blocks west of the playground. If his house is 1 block north and 2 blocks east of the playground, how far is he away from home?

   A. 1 block west and 4 blocks south

   B. 3 blocks north and 3 blocks east

   C. 6 blocks north and 5 blocks east

   D. 5 blocks north and 6 blocks west

5.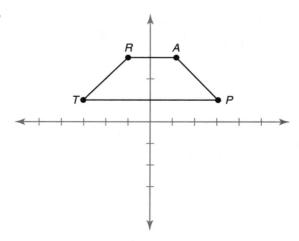

Which statement is NOT true when the figure is reflected over the *x*-axis?

A. *T'* is located at (–3, –1)

B. *R'* is located at (–3, –1)

C. *A'* is located at (1, –3)

D. *P'* is located at (3, –1)

Extended Constructed Response (Use the space provided to show your work.)

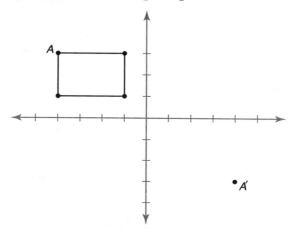

The figure has Point *A* at (–4, 3). Perform two transformations to move Point *A* to its transformed image shown on the graph.

LESSON 6: CUSTOMARY MEASUREMENT

Even though the rest of the world has adopted the metric system of measurement, the United States continues to use what is known as customary measurements. Customary measurements encompass distance, capacity, and weight. Time is universally understood. In this section, we will explore some components of the Mathematics Reference Sheet and solve some problems that deal with this system of measurement.

DISTANCE	CAPACITY	TIME	WEIGHT
12 inches = 1 foot	8 fluid ounces = 1 cup	60 seconds = 1 minute	16 ounces = 1 pound
3 feet = 1 yard	2 cups = 1 pint	60 minutes = 1 hour	2,000 pounds = 1 ton
36 inches = 1 yard	2 pints = 1 quart	24 hours = 1 day	
5,280 feet = 1 mile	4 quarts = 1 gallon	7 days = 1 week	
1,760 yards = 1 mile		12 months = 1 year	
		365 days = 1 year	

To solve the kinds of problems you will see on the NJ ASK7, you should be familiar with these conversions and be able to convert, add, and subtract them with ease.

Example 1

Mario walks $\frac{3}{4}$ of a mile to school each way. How many feet does Mario walk to school?

Solution: To solve the problem, refer to the Mathematics Reference Sheet to see that 5,280 feet = 1 mile. Since Mario walks $\frac{2}{3}$ of a mile, we can multiply $\frac{3}{4} \cdot 5,280$ to get the distance in feet.

$\frac{3}{4} \cdot 5,280 = 3,960$ feet. Mario walks 3,960 feet to school.

Example 2

Your class is competing in a team long jump against the sixth grade. Your class selects its three best long jumpers and their distances are recorded. What is the total distance jumped by all three athletes?

Susan:	2 yards	2 feet	10 inches
Manuel:	2 yards	1 foot	8 inches
Xavier:	3 yards	0 feet	7 inches

Solution: Add the distances and rename any dimensions.

Susan:	2 yards	2 feet	10 inches
Manuel:	2 yards	1 foot	8 inches
Xavier:	3 yards	0 feet	7 inches
Total:	7 yards	3 feet	25 inches

Since 3 feet = 1 yard, add 1 to the yard column. Also, 25 inches = 2 feet 1 inch. Therefore, the total distance jumped is 8 yards 2 feet 1 inch.

Example 3

Kim loves ice cream so much that she has it every night. If she eats $\frac{1}{2}$ a pint each night, how many cups of ice cream does she eat in 2 weeks?

Solution: There are 7 days in a week or 14 days in 2 weeks. Each day she eats $\frac{1}{2}$ pint.

So, $\frac{1}{2} \cdot 14 = 7$, which means Kim eats 7 pints of ice cream in 2 weeks. There are 2 cups in a pint so she eats 14 cups of ice cream in 2 weeks.

SAMPLE TEST QUESTIONS (For answers, see page 186.)

Multiple-Choice (Circle the correct answer.)

1. Suzanne is 5 feet 6 inches tall. Her friend Gail is 6 feet 2 inches tall. How many inches taller is Gail?

   A. 16 inches

   B. 12 inches

   C. 8 inches

   D. 4 inches

2. Shandra weighs 128 lb. How many ounces does Shandra weigh?

    A. Between 1,000 and 1,500 ounces

    B. Between 1,500 and 2,000 ounces

    C. Between 2,000 and 2,500 ounces

    D. Between 2,500 and 3,000 ounces

3. When Kim ran her first marathon, she ran 26.2 miles in 4 hours and 11 minutes. Her goal is to cut $1\frac{1}{4}$ hours from her time. When she meets her goal, what will be her time for running the marathon?

    A. 2 hours 4 minutes

    B. 2 hours 56 minutes

    C. 3 hours 11 minutes

    D. 3 hours 56 minutes

4. During their lunch break, Denise and Karen walked 6,600 feet each. How many miles did they walk altogether?

    A. $1\frac{1}{4}$ miles

    B. $1\frac{1}{2}$ miles

    C. $2\frac{1}{4}$ miles

    D. $2\frac{1}{2}$ miles

5. The African male elephant weighs up to 12,000 lb. How many tons does an African male elephant weigh?

    A. 6 tons

    B. 7 tons

    C. 8 tons

    D. 9 tons

Short Constructed Response

6. Dori ran $\frac{3}{4}$ mile on Monday, $1\frac{1}{8}$ miles on Tuesday, $1\frac{5}{8}$

   miles on Wednesday, and rested on Thursday.
   How far must Dori run on Friday to have run
   4 miles for the week?   _____

LESSON 7: METRIC MEASUREMENT

As noted in the previous section, many countries around the world use the metric system of measurement. We use it to some degree here in the United States. We buy soda in 2-liter bottles. Canned goods are measured in grams. Nutrition labels use metric units of measure. Next time you go to eat something, check out the nutrition label and see how many grams of protein or carbohydrates it has.

Just like the customary unit of measurement, the metric system measures distance and weight. One of the reasons why the metric system is so appealing is that it is based on a system of 10. It also uses a series of prefixes to indicate the place value of the measurement. Look at this chart.

Kilo-, k-	Hecto-, h-	Deka-, da-	Units	Deci-, d-	Centi-, c-	Milli-, m-
1,000	100	10	1	0.1	0.01	0.001

The numbers listed below the prefixes represent how many of the particular unit you have, where the units represent the unit of measure (grams, meters, liters). For example, if you were measuring something in meters, a kilometer would equal 1,000 meters. If you were measuring something in centimeters, each centimeter would

represent $\frac{1}{100}$ of a meter. An easier way of converting is to

identify the starting unit of measure and move the decimal point that many places in that direction to get the desired conversion.

$$1 \text{ kilometer} = \underline{\hspace{1cm}} \text{ meters}$$

Since meters are the desired unit and are three places to the right of kilometers, you need to move the decimal point three places to the right. 1 kilometer becomes 1,000 meters.

1 Kilometer = _____ meters

1.000

500 milliliters = _____ centiliters

Centiliters is one unit to the left of milliliters. Move the decimal point one place to the left of 500 to get the answer.

500 milliliters = 50 centiliters

500 mililiters = _____ centiliters

50.0

The units of measure in the metric system are: meters (distance), grams (mass or weight), and liters (volume).

Example 1

Kevin ran a 3-kilometer race. How many centimeters are there in 3 kilometers?

Solution: You need to convert 3 kilometers to the correct number of centimeters. Centimeters are five places to the right of kilometers. Write the number 3 and move the decimal point 5 places to the right to equal 300,000 centimeters. There are 300,000 centimeters in 3 kilometers.

Example 2

Frances and Margaret are measuring pieces of string. Frances has a 38-centimeter piece of string. Margaret has a 64-centimeter piece of string. If they put their pieces together, do they have more than 1 meter of string? Explain.

Solution: If we combine their two pieces of string, Frances and Margaret will have enough to make 1 meter. There are 100 centimeters in a meter and their combined string equals 102 centimeters. 38 cm + 64 cm = 102 cm

Example 3

1 kilogram ≈ 2.2 lb. A heavyweight boxer weighs 98 kg. How many pounds does the heavyweight boxer weigh?

Solution: Set up a proportion to solve the problem.

$$\frac{1\,kg}{2.2\,lb} = \frac{98\,kg}{x}$$
$$x = 2.2 \cdot 98$$
$$x = 215.6$$

The heavyweight boxer weighs 215.6 lb.

SAMPLE TEST QUESTIONS (For answers, see pages 186-187.)

Multiple-Choice (Circle the correct answer.)

1. Over the summer, the Anderson family travels from Eatontown, New Jersey, to Cape May, New Jersey. If the trip is approximately 95 miles and there are 1.6 kilometers per mile, how many kilometers are there from Eatontown to Cape May?

   A. 57 km

   B. 96.6 km

   C. 152 km

   D. 160 km

2. Many people go on diets to lose weight. If someone lost 14 kg in one year, how much weight did they lose?

   A. Between 15 and 20 lb

   B. Between 20 and 25 lb

   C. Between 25 and 30 lb

   D. Between 30 and 35 lb

3. If you consume an energy bar that contains 21 grams of protein, how many milligrams of protein would you consume?

   A. 210,000 mg

   B. 21,000 mg

   C. 2,100 mg

   D. 210 mg

4. After the football game on Friday, Brendan, the captain of the football team, buys eight 2-liter bottles of soda for a party at his house. Which measurement is NOT the equivalent of eight 2-liter bottles?

   A. 0.16 kL
   B. 160 dL
   C. 1,600 cL
   D. 16,000 mL

5. A moving company places some furniture in an elevator. The television cabinet weighs 24 kg. The dining room table weighs 17 kg. The bedroom set weighs 39 kg. If the two movers weigh 77 kg and 101 kg, which situation is true?

   A. The furniture weighs more than both movers.
   B. The furniture weighs more than 250 kg.
   C. The television cabinet and the dining room table weigh less than the bedroom set.
   D. The combined weight of the movers and the furniture is greater than 250 kg.

Short Constructed Response

6. How many kilometers are there in 1,600 meters? _____

Extended Constructed Response (Use the space provided to show your work.)

You and your family are driving from Newark, New Jersey, to Los Angeles, California. It is approximately 4,462 km.

▪ If your family drives 95 km/hour, how many hours will it take to reach Los Angeles?
▪ If you were to run from Newark to Los Angeles and you can run at 8 km/hour, how many hours would it take you to arrive in Los Angeles?

LESSON 8: AREA AND PERIMETER OF POLYGONS

To calculate the perimeter of a polygon, you need to add up all sides in order to get the distance around the polygon.

When calculating the area of a polygon you need to determine the amount of space inside of the polygon. Area is always expressed in terms of square units.

There are a few ways to calculate the area and perimeter of a polygon. Formulas are listed on your Mathematics Reference Sheet to help you calculate that information. However, it is important that you have a clear understanding of these concepts and can apply that information accordingly to various situations.

Take a moment to think about reasons why you would need to calculate area or perimeter. For example, if you wanted to hang border in a bedroom, you would need to measure the perimeter of the room. Calculating the perimeter would also be necessary if you planned to install fencing around your yard or a frame around a picture. You would need to calculate area if you were trying to determine how much carpeting you need in a room, how many square tiles you would need in your bathroom, or how much grass seed you would need to cover your lawn.

> *****V.I.P.*** (Very Important Point)**
>
> It is very easy to confuse area and perimeter when solving problems. Make certain that you understand the differences between each so you can solve these problems correctly.

Example 1

Calculate the area and perimeter of the following rectangle.

18 m

43 m

Solution:

Perimeter for a rectangle = 2 $(l + w)$ = 2 (43 + 18) = 2 (61) = 122 m

Area for a rectangle = $l \cdot w$ = 43 · 18 = 774 ft^2

Example 2

A right triangle has a leg that measures 5 cm, another leg that measures 12 cm, and a hypotenuse that measures 13 cm. Find its area and perimeter.

Solution:

Perimeter = 5 cm + 12 cm + 13 cm = 30 cm

$$\begin{aligned}
\text{Area} &= \frac{1}{2}bh \\
&= \frac{1}{2}(12)(5) \\
&= \frac{1}{2}(60) \\
&= 30
\end{aligned}$$

The perimeter is 30 cm and the area is 30 cm^2.

Example 3

An isosceles trapezoid has an area of 144 square feet. If its bases are 12 feet and 24 feet, what is its height?

Solution: Draw an isosceles trapezoid to help you solve the problem.

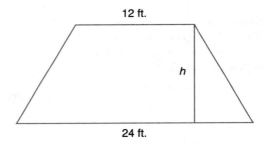

The formula to calculate the area of a trapezoid is

$$A = \frac{1}{2}h(b_1 + b_2)$$

Next, plug in the information to the formula to get the height.

$$\begin{aligned}
A &= \frac{1}{2}h(b_1 + b_2) \\
144 &= \frac{1}{2}h(12 + 24) \\
144 &= \frac{1}{2}h(36) \\
144 &= 18h \\
\frac{144}{18} &= \frac{18h}{18} \\
8 &= h
\end{aligned}$$

The height of the trapezoid is 8 feet.

SAMPLE TEST QUESTIONS (For answers, see page 187.)

Multiple-Choice (Circle the correct answer.)

1. You want wallpaper border around your bedroom. If your bedroom measures 18 feet long and 15 feet wide, how much border is needed to decorate your bedroom?

    A. 33 feet

    B. 66 feet

    C. 270 feet

    D. 540 feet

2. Jill is painting a mural at school but only has enough paint to cover 75% of the mural. If the wall where the mural is going to be painted measures 10 m long and 8 m wide, how much area will NOT be painted?

    A. 80 m^2

    B. 60 m^2

    C. 20 m^2

    D. 18 m^2

3.

    A playground in the shape of a right triangle needs to be enclosed with fencing.

    How much fencing is needed for the playground?

    A. Less than 100 feet

    B. More than 100 feet but less than 150 feet

    C. More than 200 feet but less than 250 feet

    D. More than 250 feet

4. A trapezoid has an area of 84 in.^{2}. What could be the dimensions of the trapezoid?

   A. $b_1 = 3, b_2 = 4, h = 12$

   B. $b_1 = 8, b_2 = 4, h = 14$

   C. $b_1 = 5, b_2 = 8, h = 12$

   D. $b_1 = 6, b_2 = 6, h = 12$

5. In a three-bedroom house, one bedroom measures 22 feet by 18 feet. Another bedroom measures 15 feet by 18 feet, and the master bedroom measures 33 feet by 21 feet. If a gallon of paint covers approximately 350 square feet, how many cans of paint will you need to paint the three bedrooms?

   A. 3

   B. 4

   C. 5

   D. 6

Short Constructed Response

6. Find the height of a parallelogram with an area of 72 ft^2 and a base of 18 feet. _____

Extended Constructed Response (Use the space provided to show your work.)

15 in.

40 in.

Given the rectangle above, Fred thinks that if you increase the length by 10% and decrease the width by 10%, the area will stay the same. Jeri believes that Fred is incorrect. Who is right, and explain your reasoning using complete sentences.

LESSON 9: AREA OF COMPOSITE FIGURES

Composite figures are shapes or designs that are made up of familiar figures. When calculating the area of these figures, it is important to break the figure into simple shapes where area is easy to calculate. Often times, finding the area of a composite figure requires you to find a few areas and then add them together. Keep in mind that there are usually multiple ways to find the area of a composite figure, so if such a question is asked in a constructed response format, make certain that you can clearly explain how you obtained your solution.

Example 1

A local college has a small garden near one of its buildings. The college landscaper wants to plant flowers in this space, but she is not sure of how much area she has to cover. Looking at the drawing of the garden, devise a plan and calculate its area.

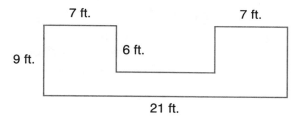

Solution: The garden looks like a rectangle with a piece of it removed. Draw a dotted line across the top to form a rectangle whose length is 21 feet and width is 9 feet. That dotted line must measure 7 feet because its other parts are congruent to the side measuring 21 feet. Since two measures are known, totaling 14 feet, 21 − 14 = 7 feet. Calculate the area of the entire rectangle, and then subtract the area enclosed by the dotted line.

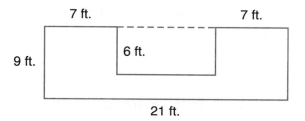

Large Rectangle:

$$\text{Area} = 21 \text{ feet} \cdot 9 \text{ feet}$$
$$= 189 \text{ feet}^2$$

Small Rectangle or Space Enclosed by the Dotted Line:

$$\text{Area} = 7 \text{ feet} \cdot 6 \text{ feet}$$
$$= 42 \text{ feet}^2$$

Area of large rectangle – Area of small rectangle = Area of garden

$$189 \text{ feet}^2 - 42 \text{ feet}^2 = 147 \text{ feet}^2$$

The area of the garden is 147 square feet.

Example 2

A building contractor wants to pour cement to form a walkway around two small seating areas in the town center. How much cement is needed to form the walkway?

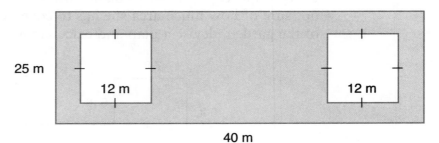

25 m   12 m   12 m

40 m

Solution: The amount of cement needed can be determined by subtracting the area of the two small seating areas from the large rectangle.

Area of Large Rectangle:

$$\text{Area} = 40 \text{ m} \cdot 25 \text{ m}$$
$$= 1{,}000 \text{ m}^2$$

Area of Small Seating Area:

$$\text{Area} = 12 \text{ m} \cdot 12 \text{ m}$$
$$= 144 \text{ m}^2$$

Because there are two small seating areas, multiply the area by 2.

Area of walkway = Area of large rectangle – 2
(Area of small seating area)

$$Area = 1{,}000 \text{ m}^2 - 2(144 \text{ m}^2)$$
$$= 1{,}000 \text{ m}^2 - 288 \text{ m}^2$$
$$= 722 \text{ m}^2$$

SAMPLE TEST QUESTIONS (For answers, see page 188.)

Multiple-Choice (Circle the correct answer.)

1.

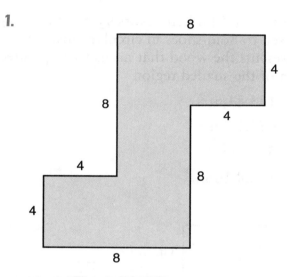

A tile, shaped like the one shown, is used in a puzzle. What is the area of this puzzle piece?

A. 64 units^2

B. 80 units^2

C. 84 units^2

D. 144 units^2

2.

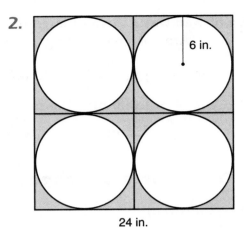

24 in.

At the Wee Children's Nursery, the following piece of furniture is used to hold shoes in circular bins. The shaded region represents the wood that needs to be painted. How much is the area of the shaded region?

A. 123.84 in.^{2}

B. 452.16 in.^{2}

C. 576 in.^{2}

D. 1,028.16 in.^{2}

3.

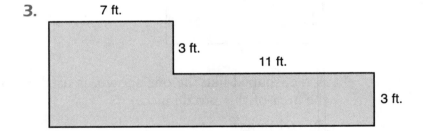

Jack visits his older brother's apartment and tries to figure the area of the living room and dining room as shown in the picture. Jack's brother wants to cover the space with laminate flooring. What is the area of the living room and dining room combined?

A. 33 ft^2

B. 42 ft^2

C. 75 ft^2

D. 108 ft^2

4.

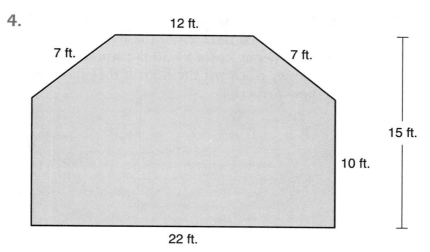

Polygonal Construction has just completed building a living room for a luxurious house. Determine how much carpeting is needed to cover the area of the living room.

A. 330 ft^2

B. 305 ft^2

C. 280 ft^2

D. 220 ft^2

Extended Constructed Response (Use the space provided to show your work.)

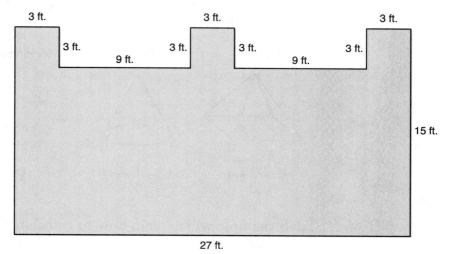

The Ask family is getting ready to move into their new home. Prior to moving, they would like to have the living room carpeted. Examine the diagram and determine the following answers:

- What is the area of the living room? Explain how you determined the area of the living room.
- If carpeting costs $4.50 per square foot, do they have enough money to carpet the floor if they have budgeted $1,500 for their new carpet?

LESSON 10: VOLUME OF PYRAMIDS, CONES, PRISMS, AND CYLINDERS

If a shape has the ability to hold something, meaning that it has capacity, then you can calculate how much it can hold by finding its volume. The following four shapes are on the Mathematics Reference Sheet.

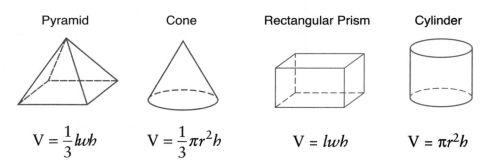

Pyramid	Cone	Rectangular Prism	Cylinder

$$V = \frac{1}{3}lwh \qquad V = \frac{1}{3}\pi r^2 h \qquad V = lwh \qquad V = \pi r^2 h$$

Each example will explain how to calculate the volume for each figure. Whenever you calculate volume, the answer is always expressed in units cubed (raised to the third power).

Example 1

Find the volume of a pyramid whose length = 4 in., width = 3 in., and height = 8 in.

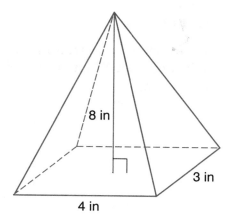

Solution:

$$\text{Volume} = \frac{1}{3}lwh$$
$$= \frac{1}{3}(4 \cdot 3 \cdot 8)$$
$$= \frac{1}{3}(96)$$
$$= 32 \text{ in.}^3$$

Example 2

You and your family are moving and have purchased boxes to pack your belongings. The storage box has a length of 30 cm, a width of 15 cm, and a height of 30 cm. What is the volume of the storage box?

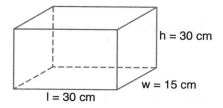

Solution:

$$\text{Volume} = lwh$$
$$= 30 \cdot 15 \cdot 30$$
$$= 13,500 \text{ cm}^3$$

The storage box can hold 13,500 cm³.

Example 3

An orange juice carton has a volume of 401.92 in³. The can has a radius of 4 in. What is the height of the orange juice can? *Note:* Use 3.14 for π.

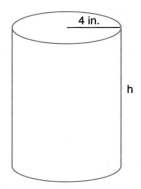

Solution: The orange juice can is a cylinder. The formula for the volume of a cylinder is Volume = $\pi r^2 h$. To find the radius, substitute all information into the formula and solve for h.

$$\text{Volume} = \pi r^2 h$$
$$401.92 = 3.14 \cdot 4^2 \cdot h$$
$$401.92 = 3.14 \cdot 16 \cdot h$$
$$401.92 = 50.24h$$
$$\frac{401.92}{50.24} = \frac{50.24h}{50.24}$$
$$8 = h$$

The height of the orange juice can is 8 inches.

SAMPLE TEST QUESTIONS (For answers, see pages 188-189.)

Multiple-Choice (Circle the correct answer.)

1. A model pyramid has a length of 18 inches and a width of 18 inches. It is twice as tall as it is wide. What is the volume of the pyramid?

    A. 324 ft³

    B. 648 ft³

    C. 3,888 ft³

    D. 11,664 ft³

2. Funnels are cone-shaped objects used to pour liquid into objects. You are pouring water into a bottle with a narrow neck and want to make sure you don't spill any water out of the bottle. You select a funnel that has a radius of 8 inches and a height of 12 inches. What is the volume of the funnel?

   A. 200.96 in.^{3}

   B. 301.44 in.^{3}

   C. 803.84 in.^{3}

   D. 2,411.52 in.^{3}

3. A tissue box cube is placed in your bathroom. If the tissue box measures 7 inches per side, what is the volume of the tissue box?

   A. 21 in.^{3}

   B. 49 in.^{3}

   C. $114\frac{1}{3}$ in.^{3}

   D. 343 in.^{3}

4. A snack can shaped like a cylinder greatly reduces the risk of chips being smashed during shipment. What is the volume of a cylindrical snack can that measures 14 inches tall and 4 inches in diameter?

   A. 175.84 in.^{3}

   B. 256.12 in.^{3}

   C. 351.68 in.^{3}

   D. 703.36 in.^{3}

Short Constructed Response

5. What is the volume of a cone that has a radius of 4 cm and a height of 6 cm? _____

Extended Constructed Response (Use the space provided to show your work.)

Find the dimensions of a rectangular prism that has a volume of 192 units^3. Show your work and explain how you obtained your answer.

LESSON 11: SURFACE AREA OF PRISMS AND CYLINDERS

Have you ever started to wrap a present and found out that you didn't have enough paper? That's because the area of the wrapping paper isn't enough to cover the surface area of the present. Simply stated, surface area is the sum of the areas for all of the faces of a prism or cylinder.

Your Mathematics Reference Sheet has the formulas for the surface area of a rectangular prism and a cylinder. They are also listed here.

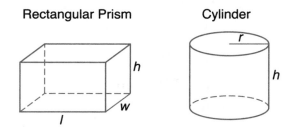

Rectangular Prism   Cylinder

$$SA = 2lw + 2wh + 2lh \quad SA = 2\pi rh + 2\pi r^2$$

Example 1

Candi is wrapping a present for her sister Lori's birthday, but is not sure she has enough wrapping paper. She has 250 in.^{2} of wrapping paper. The box has a length of 12 inches, a width of 6 inches, and a height of 3 inches. What is the surface area of the box and does Candi have enough wrapping paper for the present?

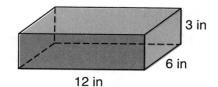

Solution: Take the dimensions of the box and substitute them into the surface area formula found on your Mathematics Reference Sheet.

$$SA = 2lw + 2wh + 2lh$$
$$= (2 \cdot 12 \cdot 6) + (2 \cdot 6 \cdot 3) + (2 \cdot 12 \cdot 3)$$
$$= 144 + 36 + 72$$
$$= 252 \text{ in.}^2$$

The box has a surface area of 252 in.^{2}, and Candi does not have enough paper to wrap Lori's present.

Example 2

A cylindrical can of potato chips has a radius of 5 cm and a height of 26 cm. What is the surface area of the can?

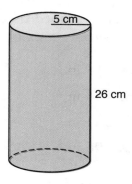

Solution: Take the dimensions of the cylinder and substitute them into the surface area formula found on your Mathematics Reference Sheet.

$$SA = 2\pi rh + 2\pi r^2$$
$$= 2(3.14)(5)(26) + 2(3.14)(5^2)$$
$$= 816.4 + 157$$
$$= 973.4 \text{ in.}^2$$

The surface area of the cylinder is 973.4 in.^{2}.

Example 3

The surface area of a rectangular prism is 208 in ^{2}. The prism's length is 4 inches; the width is 6 inches. What is the height?

Solution: Substitute the values given into the surface area formula found on your Mathematics Reference Sheet.

$$SA = 2lw + 2wh + 2lh$$
$$208 = (2 \cdot 4 \cdot 6) + (2 \cdot 6 \cdot h) + (2 \cdot 4 \cdot h)$$
$$208 = 48 + 12h + 8h$$
$$208 = 48 + 20h$$
$$208 - 48 = 48 - 48 + 20h$$
$$160 = 20h$$
$$\frac{160}{20} = \frac{20h}{20}$$
$$8 = h$$

(Remember to combine like terms. $12h + 8h = 20h$)

The height of the rectangular prism measures 8 inches.

SAMPLE TEST QUESTIONS (For answers, see page 189.)

Multiple-Choice (Circle the correct answer.)

1. A cube-shaped tissue box has a side length of 4 inches. What is the surface area of the cube?

    A. 16 in.^{2}

    B. 36 in.^{2}

    C. 64 in.^{2}

    D. 96 in.^{2}

2. Another tissue box is shaped like a rectangle. The length is 12 inches. Its width is $\frac{1}{3}$ of its length, and its height is $\frac{1}{6}$ of its length. What is the surface area of the tissue box?

    A. 64 in.^{2}

    B. 160 in.^{2}

    C. 288 in.^{2}

    D. 2,304 in.^{2}

3. A soup can has a height of 15 cm and a diameter of 16 cm. What is the surface area of the soup can?

    A. 577.76 cm²

    B. 1,155.52 cm²

    C. 3,114.88 cm²

    D. 6,430.72 cm²

4. A can of baby formula measures 14 inches tall and has a radius of $3\frac{3}{4}$ in. What is the surface area of the can?

    A. 370.9125 in.²

    B. 418.0125 in.²

    C. 604.08 in.²

    D. 835.24 in.²

5. A rectangular prism has a width of x units, a length of $2x$ units, and a height of x^2 units. If $x = 3$, what is the surface area of the rectangular prism?

    A. 18 units²

    B. 144 units²

    C. 162 units²

    D. 198 units²

Short Constructed Response

6. Find the surface area of a rectangular prism with a length of 12 cm, width of 6 cm, and height of 4 cm.   _____

7. Find the surface area of a cylinder with a height of 30 cm and a radius of 12 cm.   _____

Extended Constructed Response (Use the space provided to show your work.)

The Marketing Department of ASK Manufacturing Company has to design a logo on the surface area of their rectangular boxes. Currently, their boxes measure: length = 8 inches, width = 6 inches, and height = 4 inches. They have been asked to design a box with a smaller surface area. State the new dimensions of the box and show that the new dimensions have a smaller surface area. Make sure to show your work.

LESSON 12: SCALE DRAWINGS / DILATIONS

A dilation occurs when a figure is enlarged or reduced proportionally. A scale factor is used to multiply the x and y coordinates of each ordered pair in a figure to form the new figure. To get the new coordinates for the dilated image, multiply each coordinate by the scale factor. Let's take a look at this example and see how the dilation changes the original figure.

Example 1

A rectangle is drawn on a coordinate plane where the following points are located:

$A(-3, 2)$, $B(1, 2)$, $C(1, -1)$, and $D(-3, -1)$. What is the image of the dilation when it has a scale factor of 2?

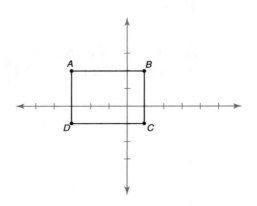

Solution: Multiply each ordered pair by 2 to get the coordinates of the new image.

$$A(-3, 2) \rightarrow A'(-6, 4)$$
$$B(1, 2) \rightarrow B'(2, 4)$$
$$C(1, -1) \rightarrow C'(2, -2)$$
$$D(-3, -1) \rightarrow D'(-6, -2)$$

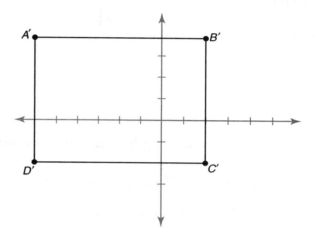

Scale drawings use a proportion to make a relationship between what a figure looks like on paper and what it looks like in reality. Model cars or blueprints are examples of how scale drawings are used.

Example 2

An architect's blueprint shows plans for a new living room. Each $\frac{1}{2}$ inch = 6 feet. How long is the actual living room if the plans show it as $2\frac{1}{2}$ inches long?

Solution: Set up a proportion to solve for the actual length of the living room.

V.I.P. (Very Important Point)

When working with dilated figures, there are changes in the perimeter and area. When calculating a new perimeter and area for any dilated figure, make certain you have the new coordinates and use the same formulas for calculating perimeter and area.

$$\frac{plan}{actual} = \frac{room}{actual}$$

$$\frac{scale}{actual} = \frac{scale}{actual}$$

$$\frac{\frac{1}{2}\,inch}{6\;feet} = \frac{2\frac{1}{2}\,inches}{x}$$

$$\frac{1}{2}x = 6 \cdot 2\frac{1}{2}$$

$$\frac{1}{2}x = 15$$

$$2 \cdot \frac{1}{2}x = 2 \cdot 15$$

$$x = 30\;feet$$

The actual length of the living room is 30 feet.

SAMPLE TEST QUESTIONS (For answers, see page 190.)

Multiple-Choice (Circle the correct answer.)

1. Rectangle *ABCD* has coordinates at *A* (–1, 5), *B* (4, 5), *C* (4, 2), and *D* (–1, 2). What are the coordinates of vertex *D* when the rectangle is dilated using a scale factor of 3?

   A. (3, 6)

   B. (–3, 6)

   C. (2, 5)

   D. (–4, –1)

2. A square has a perimeter of 8 units and an area of 4 units squared. If a scale factor of 2 is applied to the square, what is its new area?

   A. 16 units^2

   B. 8 units^2

   C. 6 units^2

   D. 4 units^2

3. On a sunny summer afternoon, a 30 foot tall tree casts a shadow that is 75 feet long. Down the street, there is a 48 foot tall tree, and it casts a shadow that is how long?

   A. 78 feet

   B. 90 feet

   C. 120 feet

   D. 123 feet

4.

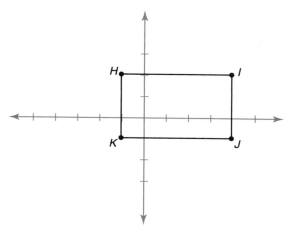

   Examine rectangle *HIJK*. If a scale factor of 2 is applied to the rectangle, what is the length of side *JK*?

   A. 4 units

   B. 6 units

   C. 8 units

   D. 10 units

5. Parallelogram *MNOP* has coordinates at *M* (0, 0), *N* (6, 8), *O* (12, 8), and *P* (6, 0). If a scale factor of $\frac{1}{2}$ is applied to the parallelogram, which statement is true.

   A. The area is reduced by $\frac{1}{2}$.

   B. Point *O* is located at (6, 8).

   C. Point *P* is located at (3, 0).

   D. Point *N* stays the same.

Short Constructed Response

6. A model car is built on a scale of $\dfrac{1 \text{ inch}}{4 \text{ feet}}$. If a car's actual length is 13 feet, how many inches long is its model? Round your answer to the nearest hundredth of an inch.

_____

Chapter 3

PATTERNS AND ALGEBRA

LESSON 1: EVALUATING ALGEBRAIC EXPRESSIONS AND EQUATIONS

In algebra, you will have to evaluate algebraic expressions or equations. The primary difference between an expression and equation is that an equation has an equal sign making the expression equal to a specific number. When you are directed to evaluate an algebraic expression, you are to determine a numerical value for that expression. Algebraic expressions or equations contain variables that can represent many values unless they are assigned a specific value.

V.I.P. (Very Important Point)

When evaluating algebraic expressions, always follow the order of operations. Simply replace the variable with the given value and calculate. Remember that a number in front of a variable is called a *coefficient*. The coefficient is always multiplied with the variable.

Example 1

Given an expression like $2a + 3b$ where $a = 5$ and $b = 9$, find the value of this algebraic expression.

Solution: Substitute the values for a and b into the expression to determine its value.

$$2a + 3b$$
$$2(5) + 3(9)$$
$$10 + 27$$
$$37$$

Example 2

Evaluate the expression $2x^2 - 4x + 9$ when $x = 6$.

Solution: Plug in 6 wherever you see an x.

$2x^2 - 4x + 9$

$2(6)^2 - 4(6) + 9$ Plug in a 6 wherever you see an x.

$72 - 24 + 9$ Follow the order of operations.

57 Solution

Example 3

Pens cost $\$.50$ and pencils cost $\$.30$. Given the equation $C = \$.50x + \$.30y$ where C = cost, x = pens, and y = pencils, calculate the cost of buying 7 pens and 9 pencils.

Solution: To evaluate the equation, $C = \$.50x + \$.30y$, plug in 7 for x and 9 for y.

$C = \$.50x + \$.30y$

$C = \$.50(7) + \$.30(9)$ Plug in 7 for x and 9 for y.

$C = \$3.50 + \$2.70$ Follow the order of operations.

$C = \$6.20$ Solution

SAMPLE TEST QUESTIONS (For answers, see pages 190–191.)

Multiple-Choice (Circle the correct answer.)

1. Evaluate $7a - 3b$ when $a = 4$ and $b = (-2)$.

   A. The answer is less than zero.

   B. The answer is equal to zero.

   C. The answer is greater than zero.

   D. The answer cannot be determined from the information given.

2. Evaluate a^3 _____ $4a^2$ when $a = 4$.

   A. $<$

   B. $=$

   C. $>$

   D. Cannot be determined

3. Evaluate $3d^2$ when $d = 5$.

   A. 30
   B. 75
   C. 225
   D. 1,225

4. Regular-size candy bars sell for $.89. King-size candy bars sell for $1.29. Let R = regular-size candy bars, K = king-size candy bars, and C = cost. Which equation could be used to evaluate a purchase of candy bars?

   A. $C = \$.89R + \$1.29K$
   B. $C = \$.89K + \$1.29R$
   C. $C = \$2.18RK$
   D. $C = \$.89 + \$1.29$

5. The cost of mailing a box is determined by the weight of the package in ounces. You can calculate the cost of shipment by using the formula $C = \$.37x$ where x represents the number of ounces. If your package weight is 14 ounces, how much will it cost to ship the package?

   A. $5.18
   B. $5.78
   C. $7.98
   D. $14.37

6. You want to make a batch of mixed nuts consisting of 2 parts peanuts, 1 part almonds, and 1 part cashews. If almonds cost $7.99 per pound, cashews cost $8.99 per pound, and peanuts cost $4.99 per pound, which expression would you use to calculate the cost of the mixture? Let A = cost of almonds per pound. Let C = cost of cashews per pound. Let P = cost of peanuts per pound.

   A. $7.99 + $8.99 + $4.99
   B. $21.97X
   C. $.25A + .25C + .5P$
   D. $\$7.99A + \$8.99C + \$4.99P$

Short Constructed Response

7. Evaluate $3a^2b$ when $a = 4$ and $b = 2$. _____

Extended Constructed Response (Use the space provided to show your work.)

On a recent math quiz, the seventh grade students in Mrs. Lopez's class were asked to evaluate the expression $4r + 3s - 2t$ when $r = 6$, $s = 4$, and $t = 3$. Paulina evaluated the expression and determined its value to be 57. Is she correct or incorrect? If Paulina's response is correct, show the work proving her answer is correct. However, if you believe that Paulina's response is incorrect, determine what she did incorrectly and show the correct way to evaluate the expression.

LESSON 2: SOLVING ALGEBRAIC EQUATIONS

Let's play a game called, "Find the Missing Number." In this game, I am going to give you an equation, and you have to determine what the missing number is as represented by X. Here we go.

Problem #1: $X + 7 = 15$
Problem #2: $X - 12 = 20$
Problem #3: $2X + 8 = 16$

I would like you to think of an equation as a seesaw or teeter-totter. The balance point is the equal sign, and each side must have an equal value in order to keep the seesaw balanced. Go back and look at Problem #1. Let's think of that problem like this.

$X + 7$ 15

To solve an equation, you have to understand a series of steps designed to help you achieve the correct value for the variable. These steps use the principle of inverse operations. Inverse operations are designed to keep the balance of an equation. So, if you are trying to solve an addition equation, you will use subtraction to solve it and vice versa. Also, if you are trying to solve a multiplication equation, you will use division to solve it and vice versa.

Keep in mind that whatever you do to one side of the equation, you MUST do to the other side. Otherwise, the equation cannot be balanced.

FOUR STEPS FOR SOLVING AN EQUATION

Follow these four steps, and you can solve any equation.

1. Isolate. Identify the variable by putting a box around it, circling it, bolding it, or any other visual clue. This will allow you to identify the inverse operation you will need to use.

2. Inverse. In this step, use the inverse operations needed to get the variable by itself.

3. Solve. Do the math! Perform the necessary calculations to get the value for the variable. Know that, depending on the problem, you may have to work backwards in order to obtain an answer.

4. Check. Make sure you check the work. To do this, plug your answer into the original equation and determine that both sides of the equation are the same.

Example 1

$$X + 19 = 45$$

$\boxed{X} + 19 = 45$ Isolate.

$\underline{\quad\quad - 19 \quad -19}$ Inverse—Subtract 19 from each side.

$X = 26$ Solve.

$26 + 19 = 45$ Check.

$45 = 45$

Example 2

$$9X = 144$$

$$9\boxed{X} = 144 \qquad \text{Isolate.}$$

$$\frac{9\boxed{X}}{9} = \frac{144}{9} \qquad \text{Inverse—Divide each side by 9.}$$

$$X = 16 \qquad \text{Solve.}$$

$$9(16) = 144 \qquad \text{Check.}$$

$$144 = 144$$

Example 3

$$3X - 21 = 63$$

$$3\boxed{X} - 21 = 63 \qquad \text{Isolate.}$$

$$\underline{+ 21 \quad + 21} \qquad \text{Inverse—The first inverse is to add 21 to both}$$

$$3X = 84 \qquad \text{sides.}$$

$$\frac{3X}{3} = \frac{84}{3} \qquad \text{Inverse—Divide both sides by 3 to get the value}$$
$$\text{for } X \text{ by itself.}$$

$$X = 28 \qquad \text{Solve.}$$

$$3(28) - 21 = 63$$

$$84 - 21 = 63$$

$$63 = 63 \qquad \text{Check.}$$

SAMPLE TEST QUESTIONS (For answers, see page 191.)

Multiple-Choice (Circle the correct answer.)

1. Which step would you perform to both sides of the equation FIRST to solve for x in the equation $2x + 7 = 15$?

    A. Divide by 2

    B. Subtract 7

    C. Add 7

    D. Multiply by 2

2. What is the value for x in the equation $6(x + 2) = 54$?

   A. 4

   B. 5

   C. 6

   D. 7

3. A CD costs $12. If you spent $96 on CDs, how many CDs did you purchase?

   A. 7

   B. 8

   C. 9

   D. 10

4. For a family vacation, you need to rent a car. The car rents for $22.50 per day with a fifteen-cent charge per mile driven. Which equation would you use to determine the total cost for renting the car? Let C = Cost and m = miles driven.

   A. $C = \$22.50m + \$.15$

   B. $C = \$22.65m$

   C. $C = \$22.50 + \$.15m$

   D. $C = \$22.50/\$.15$

5. Refer to question 4. Since the car rents for $22.50 per day and you are charged $.15 per mile, how many miles could you drive and not spend over $50.00?

   A. 150

   B. 183

   C. 184

   D. 333

6. Solve for x: $3(x + 7) = 48$.

   A. $\dfrac{11}{3}$

   B. 9

   C. 13

   D. 16

(handwritten work:
$3X + 21 = 48$
$-21 = -21$
$3X = 27$
$X = 9$ *)*

Short Constructed Response

7. What is the value of x in the equation $4x + 17 = 101$? _____

Extended Constructed Response (Use the space provided to show your work.)

You received a $20 gift card for downloading songs to your media player. You are charged $1.25 for each song.

- Write an equation to model the number of songs you can download to your media player.
- Solve for X to determine the number of songs you can download with your $20 gift card.
- Write an equation and determine the number of songs you could download if you were given a $50 gift card.

LESSON 3: DETERMINING THE RULE FOR A PATTERN

When looking at a table of values, there are two columns. The first column is an input column and the second is an output column. An output is always determined by a rule that is given in the problem. Let's look at this example.

Input, X	Output, $4X$
1	4
2	8
3	12
4	16

As you can see, the output rule is $4X$. That means you must multiply whatever value is given in the input column by 4 to get the output.

But what would happen if you had the output values but didn't know the rule? Examine the following table and notice that whatever value is given for X (input) it is substituted into a rule to get an output value.

Input, X	Output, ???
1	7
2	9
3	11
4	13

You can look at the output values and see that when X equals 1, the function is equal to 7. When X equals 2, the function is equal to 9 and so on. But, how can you discover what the output rule is?

When the input value increases by 1, you can find the common difference in the corresponding output values. Knowing the common difference will help you find the rule. To discover the rule, you want to examine the relationship between the input and output values. Since the input values increase by 1, you can examine the output values to find the common difference.

7	9	11	13
First Term	Second Term	Third Term	Fourth Term

$$+2 \qquad +2 \qquad +2$$

Notice that the common difference between each term is +2. Additionally, if the pattern dictates that you must keep increasing the previous number by 2, what do you think the "zero term" is? To find the Zero Term, you have to work backwards. Instead of adding 2, subtract 2 from the first term to find the Zero Term, which is 5. Once you know the common difference and the zero term, you can place that information into a formula to get the rule.

> *****V.I.P.*** (Very Important Point)**
>
> When the common difference is the same, as in the example above, we say that the relationship is linear. A linear relationship simply means that when the input value increases by 1, the output value increases by the same amount each time.

5	7	9	11	13
Zero Term	First Term	Second Term	Third Term	Fourth Term

+2   +2   +2   +2

(Common Difference)X + Zero Term → Rule: $2X + 5$

Example 1

Examine the table and determine the rule.

Input, X	Output, ???
1	–2
2	1
3	4
4	7

> *****V.I.P.*** (Very Important Point)**
>
> Some problems ask you to find the nth term. The nth term is an algebraic representation for an input value. So, rather than using X, the problem just uses the variable n. When you find the nth term, you can then define the output in terms of n. When you find the output in terms of n, you have found the rule.

Solution: The input values increase by 1 and the output values increase by 3. That means the common difference is 3. To find the Zero Term, subtract 3 from –2 to equal –5. Therefore, the rule is $3X – 5$.

SAMPLE TEST QUESTIONS (For answers, see pages 191–192.)

Multiple-Choice (Circle the correct answer.)

1. Each day, Joe earns five dollars more than he did on the previous day. Which expression correctly models the amount of money he would earn on the nth day?

   A. $5n$

   B. $5 + n$

   C. $n \div 5$

   D. $5n + 5$

2. Malachi is doing crunches to strengthen his abs. The first week he does 20 crunches each day. Each week, he will increase the number of crunches he does each day by 20. Which rule represents the number of crunches he does during the nth week?

   A. $20n$

   B. $20(n - 1)$

   C. $20 + n$

   D. 20^n

3. Which expression models the relationship in the table?

X	???
1	−7
2	−5
3	−3
7	5

   A. $2X - 9$

   B. $-X - 9$

   C. $3X - 9$

   D. $X + 4$

4. At a local movie theater, you are in charge of recording ticket sales. An adult ticket costs $9.50 while a child's ticket costs $6.00. Which expression could be used to determine the number of ticket sales? (Let A = adult ticket and C = child's ticket.)

   A. $15.50A

   B. $6.00A + $9.50C

   C. $3.50(A + C)

   D. $9.50A + $6.00C

5. A local restaurant has square tables that seat four people. In order to accommodate larger parties, the restaurant joins these tables together on one side. How many people can be seated if there are seven square tables?

   A. 10

   B. 12

   C. 14

   D. 16

6. A garden measures 12' by 15'. If you double the dimensions of the garden, what happens to its area?

   A. The area stays the same.

   B. The area doubles.

   C. The area quadruples.

   D. The area increase by 8 times its original amount.

Short Constructed Response

7. Training for a 5-mile race, you begin a running program in which you run one mile the first week and increase the distance by a half-mile each week thereafter. How many miles will you run during the ninth week? _____

Extended Constructed Response (Use the space provided to show your work.)

Your cell phone plan charges $30.00 a month for 250 minutes plus ten cents for each additional minute.

- Make a table showing the cost of using 250, 300, 350, and 400 minutes.
- Write a formula that can be used to calculate the cost of your cell phone bill when talking for m minutes each month.
- An alternate plan costs $50.00 a month for 500 minutes. Will that save you money? Explain why or why not.

LESSON 4: TYPES OF PATTERNS AND SEQUENCES

People like patterns. It's really that simple. By this I mean that people like to see things follow a certain rule or progression that is consistent. Have you ever seen a pretty necklace or bracelet? Maybe you made a friendship bracelet and used certain colors. You probably didn't just throw the colors together but placed them on the string in an orderly manner.

Imagine that you are going to make a friendship bracelet and you have three colored beads. You are going to use the colors red, white, and blue. You take your string and begin to place the beads on it. You place the red bead first, followed by the white bead, and then the blue bead. Once those three beads are on the string, you repeat the process. Red, White, Blue, Red, White, Blue, ... The pattern continues until you can't place any more beads on the string. Your bracelet will need 40 beads. As you continue the pattern, which bead will be the 40th on the string? How could you determine which colored bead will be in the 40th position?

First, you might actually take the beads and line them up.

1	2	3	4	5	6	7	8	9	10	11	12	13	14	15	16	17	18	19	20
R	W	B	R	W	B	R	W	B	R	W	B	R	W	B	R	W	B	R	W

21	22	23	24	25	26	27	28	29	30	31	32	33	34	35	36	37	38	39	40
B	R	W	B	R	W	B	R	W	B	R	W	B	R	W	B	R	W	B	R

In the table, we listed every single bead and color to find that the red bead will be the 40th bead in the bracelet. But, could there have been an easier way to solve the problem?

There are three beads (red, white, blue), which are continually repeated until the bracelet is complete. The bracelet will have 40 beads. If you divide 40 by 3 you will get an answer of 13 with a remainder of 1. The answer means that you will have 13 complete sets of beads and then 1 additional bead. Since red is the next bead after a complete set, the 40th bead will be red.

There are two types of sequences in mathematics. An *arithmetic sequence* is a series of numbers that increases by the same amount each time. Think of an arithmetic sequence as counting by a certain number, like counting by 2s. 2, 4, 6, 8, 10, etc. is an example of an arithmetic sequence. Repeated counting by a certain number is also called *skip counting*. Sometimes a sequence can start with a particular number and then increase or decrease by the same amount each time. The following are some examples of arithmetic sequences.

5, 10, 15, 20, 25, …	This sequence increases by 5.
9, 18, 27, 36, 45, …	This sequence increases by 9.
24, 21, 18, 15, 12, …	This sequences increases by –2.

A *geometric sequence* is a series of numbers in which the next number is found by multiplying by the same number each time. For example, if your sequence starts with 5 and you double the next number (multiply by 2), the sequence would be 5, 10, 20, 40, 80, etc. In contrast, an arithmetic sequence has a common difference between each number. A geometric sequence does not have that. You can see that the differences between the numbers are getting much larger as the sequence progresses.

> *****V.I.P.***** (Very Important Point)
>
> The **Fibonacci Sequence** is a famous mathematical sequence that is neither an arithmetic or geometric sequence. It begins with 1, 1, and each number thereafter is formed as the sum of the previous two numbers. The Fibonacci Sequence is 1, 1, 2, 3, 5, 8, 13 . . . .

Here are some more examples of a geometric sequence.

4, 12, 36, 108, 324, ...   Each number is multiplied by 3.

3, 30, 300, 3000, 30000, ...   Each number is multiplied by 10.

500, 250, 125, 62.5, 31.25   Each number is multiplied by $\frac{1}{2}$.

Sometimes, we might see a sequence that doesn't follow an arithmetic or geometric progression. These sequences can be a little difficult to solve as adding or multiplying by the same number each time doesn't find the next number. Here are a few examples of a sequence which we would classify as neither arithmetic nor geometric.

1, 2, 4, 7, 11, ...   Add 1, add 2, then add 3, and so on.

9, 7, 12, 10, 15, ...   Add –2, add 5, add –2, add 5, and so on.

1, 1, 2, 6, 24, ...   Multiply by 1, multiply by 2, multiply by 3, multiply by 4, and so on.

SAMPLE TEST QUESTIONS (For answers, see pages 192–193.)

Multiple-Choice (Circle the correct answer.)

1. Joaquin is making a beaded necklace. He places a green bead, followed by a yellow bead, then a blue bead, and a red bead. He repeats this process until the necklace is completed. Which bead will be the 50th bead?

   A. Green
   B. Yellow
   C. Blue
   D. Red

2. Sue runs 3 miles her first week and then increases her distance by $1\frac{1}{2}$ miles each week. How many miles will she run during her seventh week?

   A. 4.5 miles
   B. 9 miles
   C. 10.5 miles
   D. 12 miles

3. Research studies have shown that healthy weight loss occurs when people exercise regularly and limit their daily caloric intake. If a person currently consumes 3,500 calories per day and wants to cut back their daily amount by 250 calories each week, how many weeks will it take to reduce their daily caloric intake in half?

   A. 5
   B. 6
   C. 7
   D. 8

4. A farmer wants to increase the amount of corn his farm produces. Currently, his farm yields 5,000 ears of corn per season. Each year he plans to double the amount of corn his farm will yield. If this occurs, how many ears of corn will his farm yield during the fourth year?

   A. 160,000
   B. 80,000
   C. 40,000
   D. 20,000

5. A local bus leaves the station at 6:00 A.M. and returns to pick up new passengers every 40 minutes. Providing there are no delays and the bus maintains its schedule, what time could the bus NOT return to the station?

   A. 7:20 A.M
   B. 8:00 A.M.
   C. 8:20 A.M.
   D. 8:40 A.M.

6. Carly and Emily are playing with some blocks. They take the blocks and begin making a pattern with them. How many blocks will they place in the fifth arrangement?

    A. 56

    B. 42

    C. 30

    D. 20

7. Carly and Emily are now playing with circles and have designed another pattern.

Which arrangement will consist of 64 circles?

    A. 8

    B. 7

    C. 6

    D. 5

8. While training for a race, Dana runs 2 minutes a day during the first week, 3 minutes a day during the second week, and 5 minutes a day during the third week. During the fourth week, she runs for 8 minutes a day and then 13 minutes a day during the fifth week. If she continues this training pattern, how long will she run for each day during the seventh week?

    A. 18

    B. 21

    C. 23

    D. 34

Short Constructed Response

9. You are given one dollar on the first day, two dollars on the second day, and four dollars on the third day. If the pattern continues, how many dollars will you be given on the eighth day? _____

Extended Constructed Response (Use the space provided to show your work.)

▪ Draw a graph to show the relationship from the following table.

X	1	2	3	4
Y	1	4	9	16

▪ What would be the corresponding y values when x is 5, 6, and 7?

▪ Would you describe this relationship as linear? Explain why or why not.

LESSON 5: WRITING AN EXPRESSION OR EQUATION

Being able to communicate mathematically means that you can take a phrase or sentence and translate it into a mathematical expression or equation. Mathematics, in many ways, is its own language because certain words indicate particular operations or symbols. If you have ever studied a foreign language or are studying one now, you can understand how important it is to translate the meaning correctly and accurately. It is very important as you learn more about mathematics that you master this skill. Translating text into expressions and equations is very useful in algebra and more complex problem solving.

There are four mathematical operations. Listed are a series of words that have the same meaning as the operation or are associated with it.

Addition	Subtraction	Multiplication	Division
increase by	decreased by	factor	divided by
more than	less than	times	separated
plus	minus	multiplied by	quotient
total	subtract	of	
in all	difference	product	
sum			

When given a phrase or sentence, look for key words to indicate the operation and place the numbers and/or variables accordingly. Let's look at some examples.

Example 1

Write an expression for the phrase, "15 more than a number x."

Solution: The key words here are "more than," which indicate addition. Since it is 15 more than, the 15 has to come after the addition sign. The correct expression is $x + 15$.

Example 2

Write an equation for "3 less than the product of x and 4 is 17".

Solution: The key words here are "less than" and "product." The operations in this equation are subtraction and multiplication. The product of x and 4 is written as $4x$. 3 less than would be written as -3. The word "is" indicates an equal sign when writing an equation. Therefore, the correct equation is $4x - 3 = 17$.

Example 3

Write an expression for the phrase, "8 less than a number n."

Solution: "Less than" means subtraction but where do you place the 8 and the n? You have two options.

Option A: $8 - n$
Option B: $n - 8$

Option B is the correct expression because it is 8 less than a number n. You are taking away 8 from n, not taking away n from 8.

You can also check by substituting a value for n to see if it makes sense. If you let $n = 10$, what would be 8 less than 10? It would be 2. Now, which option, when $n = 10$, gives you a value of 2?

Option A: $8 - 10 = (-2)$
Option B: $10 - 8 = 2$

The correct expression is Option B, $n - 8$.

SAMPLE TEST QUESTIONS (For answers, see page 193.)

Multiple-Choice (Circle the correct answer.)

1. A shipping company charges $.89 per ounce plus the cost of a box. If the box costs $2.35, what is the equation used to calculate shipping a box that weighs x ounces?

   A. Cost = 3.24x$

   B. Cost = $2.35 + $.89

   C. Cost = 2.35x$ + $.89

   D. Cost = $.89$x$ + $2.35

2. Which equation is correctly translated as "15 less than twice a number *n* is 42."

    A. $15 - 2n = 42$

    B. $2n - 15 = 42$

    C. $2(n - 15) = 42$

    D. $2n + 15 = 42$

3. A large order of french fries costs $.75 more than a small order of french fries. Which expression could be used to calculate the cost of a large order?

    A. $.75x

    B. $.75 - x$

    C. $x + $.75

    D. $.75 / x$

4. Francesca is five years less than twice her sister Kathleen's age. Let *K* = Kathleen's age. Which expression could be used to express Francesca's age?

    A. $2K - 5$

    B. $2K + 5$

    C. $K - 2$

    D. $5 - 2K$

5. It is approximately 157 miles from Oakland, New Jersey, to Cape May, New Jersey. Which equation would you use to determine how long it will take you to travel from Oakland to Cape May if you drive an average of 65 miles per hour?

    A. $D = 65 + 157$

    B. $D = 157 - 65$

    C. $D = 157 \div 65$

    D. $D = 157 (65)$

LESSON 6: WHAT HAPPENS WHEN ... ?

Part of being a good mathematics student is to understand how variables can change within an equation. Examine the equation $x + y = 5$. Now, by looking at this equation, you might be able to determine that there really isn't one specific answer for each variable. What do you think happens to y, when x increases?

To answer this question, you have to imagine that the values for x could start at a low number and then gradually increase. In other words, if $x = 1$, y must equal 4 to make the equation true. But if $x = 2$, then y must equal 3. If $x = 3$, $y = 2$. So, as x increases, y must decrease in order to make the equation true. If you have difficulty trying to solve a problem like this, go back to an earlier section where we looked at an equation as being like a seesaw and use that idea to help you keep the equation balanced.

Sometimes, you may be asked questions about equations, functions, or types of "real-life" situations where you will have to use your algebraic reasoning in order to solve the problem. As a suggestion, make sure you understand what the question is asking you to do; make a table or write down some numbers which relate to the problem, in order to visually see the results before answering the question.

Example 1

Given the equation $2x + 3y = 30$, what must happen to y as x increases?

Solution: Think of possible values for x and y that will make the equation true. Create a table of values to help you see the relationship.

x	y	$2x + 3y = 30$
0	10	$2(0) + 3(10) = 30$
1.5	9	$2(1.5) + 3(9) = 30$
3	8	$2(3) + 3(8) = 30$

As x gets bigger, y must get smaller to keep the equation equal to 30.

Example 2

Karina is working at a local supermarket earning $7.50 per hour and her friend Lauren works at a video store earning $9.00 an hour.

■ Write an expression that represents the amount of money Karina earns during a week and the amount of money Lauren earns during a week.

■ How many complete hours must each employee work to earn the same amount of money?

■ Karina gets a raise to $8.00 an hour. How many hours must she work to have the same salary as Lauren?

Solution: To write an expression, you need to assign a variable for the number of hours each person works. Let h = the number of hours. Then, Karina earns $7.50h$ and Lauren earns $9.00h$.

■ You can construct a table of values to find when each person will make the same amount of money.

Hour, h	Karina = $7.50h$	Lauren = $9.00h$
1	$7.50	$9.00
2	$15.00	$18.00
3	$22.50	$27.00
4	$30.00	$36.00
5	$37.50	**$45.00**
6	**$45.00**	$54.00

When Karina works 6 hours she will earn $45.00, which is the same amount of money Lauren earns when she works 5 hours.

■ You can construct another table to show that when Karina gets a raise, she will earn the same amount of money as Lauren after working 9 hours, when Lauren works 8.

SAMPLE TEST QUESTIONS (For answers, see page 193-194.)

Multiple-Choice (Circle the correct answer.)

1. Given the equation $a + b = 7$, what must happen to a as b decreases?

    A. a must get smaller.

    B. a must get larger.

    C. a stays the same.

    D. There is not enough information to determine the answer.

2. Given the equation $3x + 2y = 12$, what happens when $0 < x < 6$?

   A. As x increases, y must decrease.

   B. As y increases, x must decrease.

   C. x and y are always the same value.

   D. There is not enough information to determine the answer.

3. Kathy is paid $8.00 an hour working at a local sporting goods store. Her friend Tom works at the competitor's sporting goods store earning $250 per week. Which statement is always true?

   A. Kathy will never earn as much as Tom in one week.

   B. Tom will never earn as much as Kathy in one week.

   C. Kathy earns more than Tom if she works 30 hours in one week.

   D. Tom will earn more than Kathy if Kathy works 30 hours in one week.

4. What happens to Yvette's salary when she is paid $3.00 on day 1, $6.00 on day 2, $12.00 on day 3? If the pattern continues for the rest of the week, how much will Yvette earn on day 7?

   A. Almost $300.00

   B. Almost $200.00

   C. Almost $100.00

   D. Almost $50.00

5. A scientist measures 1,000 bacteria cells in an experiment at 9:00 A.M. The bacteria are expected to double every thirty minutes. How many bacteria cells will there be at 10:30 A.M.?

   A. There will be 2,000 bacteria cells.

   B. There will be 3,000 bacteria cells.

   C. There will be 4,000 bacteria cells.

   D. There will be 8,000 bacteria cells.

Extended Constructed Response (Use the space provided to show your work.)

Explain whether it is better to be paid $7.50 an hour or $300 per week? *Note:* When someone works a full week, they typically work 40 hours.

LESSON 7: INEQUALITIES

We spend so much time in mathematics trying to find out what something equals, that when teachers ask you to find something that is NOT equal to something, that can get confusing! Imagine asking someone what 2 + 2 does NOT equal!

Inequalities are situations in mathematics where a range of values can satisfy a problem. For example, if I asked you how much money you have and you told me that you have more than $10, I wouldn't know exactly how much money you have, but I would at least know what you don't have.

There are five inequality symbols that you must know. They are:

Symbol	Meaning
$<$	Less than
$>$	Greater than
$\leq$	Less than or equal to
$\geq$	Greater than or equal to
$\neq$	Not equal to

Once you understand these symbols, you can use them to express many mathematical relationships. If you want to say that you have more than $10 and the variable m is going to represent the amount of money you have, you would write $m > \$10$.

Let's try writing a few inequalities.

Example 1

Monique is at least 25 years old.

Solution: Because Monique can be 25 years of age or older, the inequality is $M \geq 25$.

Example 2

I predict that Liesl will invite no more than 20 people to her party.

Solution: The prediction is that Liesl will invite a maximum of 20 people to her party. Therefore, the inequality is $L \leq 20$.

Example 3

You tell your parent that you scored at least a 70 on your most recent math test, but no more than 92.

Solution: You've told your parent that there is a range of values for your most recent math test. You can use two inequality symbols to express your test score by writing $70 \leq T \leq 92$. This is referred to as a compound inequality. A compound inequality uses two inequality symbols and expresses a range of values between two numbers.

V.I.P. (Very Important Point)

When answering a multiple-choice question on inequalities, the graphs will pretty much look the same with the exception of the circle and the placement of where the line is drawn to show the solution. As a suggestion, you may want to graph the solution first on your own and then match your number line with the correct answer choice. This way, you can possibly eliminate a careless mistake when solving one of these problems.

Example 4

Write a compound inequality to show that the temperature in Paterson, New Jersey, was between 58°F and 82°F.

Solution: You must be able to express that the temperature is greater than 58°F and less than 82°F. When writing the compound inequality, the variable that represents the temperature during the day, needs to go in the middle. The compound inequality is $58°F \leq x \leq 82°F$.

The table models what kinds of circles to use when graphing an inequality on a number line. Remember that an open circle means that the number is *not* part of the solution, but a shaded number means that it *is* part of the solution.

Symbol	Meaning	Circle
<	Less than	Open
>	Greater than	Open
≤	Less than or equal to	Shaded
≥	Greater than or equal to	Shaded
≠	Not equal to	Open

Example 5

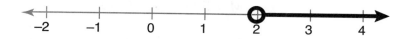

Examine the number line above and notice that there is an open circle at 2 and the number line is shaded to the right. This means that the answer is greater than 2 but does not include 2.

SAMPLE TEST QUESTIONS (For answers, see page 194.)

Multiple-Choice (Circle the correct answer.)

1. Campbell tells Marisol that she owes her brother at least $30.00. Which inequality best represents how much Campbell owes her brother?

   A. $x < \$30$

   B. $x > \$30$

   C. $x \leq \$30$

   D. $x \geq \$30$

2. While traveling on summer vacation, you and your family drive from Morristown, New Jersey, to Orlando, Florida. You estimate that you and your family will drive between 1,000 and 1,200 miles to arrive in Orlando. Which inequality best represents the distance traveled?

   A. $1,000 > x > 1,200$

   B. $1,000 < x < 1,200$

   C. $1,000 \geq x \geq 1,200$

   D. $1,000 \leq x \leq 1,200$

3. Will and Seth sell lemonade to buy a new video game. They need to earn at least $50.00. They will sell glasses of lemonade for $.50 and determine that expenses (cups, lemons, drink mix, etc.) will cost $10.00. Which inequality could they use to solve for how many cups they must sell in order to have enough money to buy the video game? (Let M = money and C = cups sold.)

   A. $M \geq \$.50C - \$10.00$

   B. $M \geq \$.50C + \$10.00$

   C. $M \leq \$.50C - \$10.00$

   D. $M > \$.50C + \$10.00$

4. The New Jersey Devils scored at least 5 goals in their last hockey game. Which graph correctly models how many goals were scored?

   A.

   B.

   C.

   D.

5. Ken and John are bowling a doubles tournament. During the first game, their combined score was 445. Which inequality best shows what they must bowl in their second game to have a total of at least 800?

   A. $x > 335$

   B. $x < 355$

   C. $x \geq 355$

   D. $x \leq 355$

DATA ANALYSIS, PROBABLITY, AND DISCRETE MATHEMATICS

LESSON 1: MEASURES OF CENTRAL TENDENCY (MEAN, MEDIAN, AND MODE)

Have you ever noticed how some athletes are better than others? Perhaps you have noticed that some people in your class score better than others on tests. Maybe you have noticed that you sometimes spend more during one week than another. Regardless of the situation, over a period of time we could express each situation by using an average.

Think about your grades for a particular class. What was your grade? That grade reflects the average or mean score you achieved on all of your tests, quizzes, etc. Let's say that your average was an 85. If you took five tests, that 85 average simply means that your score was basically an 85 on all tests. Granted some test scores might have been higher or lower than that, but when we add up all of your tests and divide by 5 (because there were 5 tests in the marking period), the answer is 85.

There are three measures of central tendency and they are mean, median, and mode. Each of these measures represents a certain part of a set of data. These definitions can be easily confused, so it is very important that you know how to calculate each measure. Check out this V.I.P.!

Measure	Definition	Study Hint
Mean	To calculate the mean, add all the numbers in the data set and divide by how many there are.	Also known as the average.
Median	To calculate the median, list the numbers in the data set in order from least to greatest and find the middle number. If the amount of numbers in the data set is even, take the average of the two middle numbers.	Median sounds like "middle." So, when you are finding the median, you are finding the middle number.
Mode	To determine the mode, find the number that appears most often in a set of data.	Mode sounds like "most."

Let's try a few examples to see how to determine the measures of central tendency.

Example 1

Shelly is watching how much she spends each week. Each week she writes how much money she has spent and compares it with the other weeks. She is hoping to save enough money so she can purchase a car when she is old enough to drive. Here are the amounts of money she spent over the past five weeks.

$8.75 $11.25 $15.50
$9.00 $11.25

Determine the mean, median, and mode for this set of data.

Solution: To calculate the mean, add up the five amounts and divide by 5.

$8.75 + $11.25 + $15.50 + $9.00 + $11.25 = $55.75 ÷ 5 = $11.15

To calculate the median, order the numbers from least to greatest and determine the middle number.

$8.75 $9.00 $11.25 $11.25 $15.50

The median is $11.25.

To calculate the mode, find the number that appears the most frequently.

The number that appears the most in the set of data is $11.25.

Example 2

You have just completed the first three weeks of your bowling league and you want to tell people how you are doing. Examine the following nine bowling scores and describe whether you would use the mean, median, or mode to best represent your bowling.

145   139   167   158   124   139   186   162

Solution: Rank the scores in order from least to greatest.

124   139   139   145   158   162   167   186

V.I.P. (Very Important Point)

When using a calculator to solve this problem make certain to press the "=" key after you have entered all the numbers to get the sum before dividing. Because scientific calculators follow the order of operations, if you do not press the = after entering all the numbers, the calculator will only divide the last number as opposed to the whole set.

Calculate the mean (round answer to the nearest tenth):

$$1220 \div 8 = 152.5.$$

Calculate the median: The numbers have been ranked in order from least to greatest, but since there is an even amount of scores, the two middle numbers are 145 and 158. To calculate the median, you must find the average of those two scores (303) and divide by 2 to get 151.5.

Calculate the mode: The most frequent number that appears in these scores is 139.

The measure of central tendency that best represents your bowling scores is the mean. Even though the median is only 1 pin less, you wouldn't want to use the mode to represent your bowling because it is significantly lower than the other two measures of central tendency.

SAMPLE TEST QUESTIONS (For answers, see pages 194–195.)

Multiple-Choice (Circle the correct answer.)

1. Michael's bowling scores for a recent tournament are as follows:

    {218, 197, 225, 246, 176, 204, 197, 257}

    What is his median score?

    A. 197

    B. 204

    C. 211

    D. 218

2. A seventh grade science class in Cape May decides to monitor the outside temperature for five days. The temperatures are 72°F, 67°F, 74°F, 79°F, and 75°F. What is the mean temperature for these five days?

    A. 73.4°F

    B. 74°F

    C. 75°F

    D. 76.2°F

3. Kimberly wants to achieve a 90 average in her math class. She has taken four tests in the marking period and her grades are 87, 94, 82, and 88. Which statement best describes what she needs on her fifth test to achieve a 90 average?

   A. Kimberly must score a 90 on her next test to achieve a 90 average.

   B. Kimberly already has a 90 average.

   C. Kimberly must score a 99 on her next test to achieve a 90 average.

   D. Kimberly cannot obtain a 90 average for this marking period.

4. What number must be added to this data set in order to have a mean of 5?

   $$\{7, 3, 5, 4, 8, ?\}$$

   A. 10

   B. 5

   C. 4

   D. 3

5. The Challenger basketball team scores are 48, 39, 43, 55, 52, and 39.

   In their next game, they score 46 points. What measure of central tendency will NOT change?

   A. Mean

   B. Median

   C. Mode

   D. They all change.

6. Brandon is calculating his mathematics average. He has had 7 tests this marking period, and his grades are recorded in the chart.

Test 1	Test 2	Test 3	Test 4	Test 5	Test 6	Test 7
98	87	90	78	83	95	99

What is Brandon's average?

A. 99

B. 98

C. 90

D. 78

Short Constructed Response

7. What is the median for the following set of data?

{18, 25, 29, 31, 33, 40}   _____

Extended Constructed Response (Use the space provided to show your work.)

The manager of a local ice cream parlor tells a new employee that the average salary of the employees is $7.50 an hour. Five employees' salaries are reported as $8.00, $7.00, $7.25, $7.50, and $8.00. Does the manager accurately report the average salary or not? Explain your answer and show your work to support your explanation.

LESSON 2: PRESENTING AND INTERPRETING DATA IN TABLES AND GRAPHS

Tables and graphs are excellent ways to present information on a particular topic. Many newspapers and magazines use tables and graphs to report information in ways that appeal to their readers. It is always important for people to correctly interpret and understand the information presented in this form.

A local video store reports monthly DVD rentals by using the following graph.

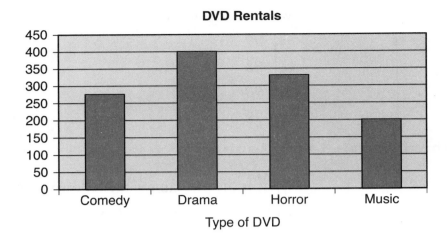

DVD Rentals

Type of DVD

Looking at the bar graph, you can easily see that Drama is the most rented type of video. Which type of video was rented half as much as Drama? You can see by the graph that Music videos are rented half as much. You can tell that the bar representing Music appears to be half as large as Drama, but you can be certain by the scale on the left side of the graph showing that Music had 200 rentals and Drama had 400 rentals.

Example 1

500 students in your school were surveyed to learn what they eat for breakfast. The results are presented in the circle graph.

What Students Eat for Breakfast

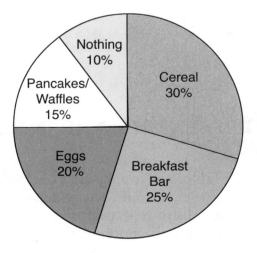

Based on the graph, which breakfast food did 75 students eat?

V.I.P. (Very Important Point)

Always remember that using estimation techniques can greatly save you time in answering certain questions.

Solution: Notice that there are no number of students listed, but percentages. In order to obtain the correct answer, you must look at the graph, take the percentage and multiply it by 500 (the amount of students surveyed) to determine which breakfast food was eaten by 75 students.

Cereal:	$30\% \times 500 = 150$
Breakfast bar:	$25\% \times 500 = 125$
Eggs:	$20\% \times 500 = 100$
Pancakes/waffles:	$15\% \times 500 = 75$
Nothing:	$10\% \times 500 = 50$

A frequency table is used to present information that is based on a range of values.

Example 2

The frequency table displays the amount of money earned per hour by employees on a particular job.

Salary Per Hour	Frequency
$6.00–$6.99	12
$7.00–$7.99	9
$8.00–$8.99	15
$9.00–$9.99	5

Which wage is earned by the most people?

Solution: The largest frequency is 15 and since they earn between $8.00 and $8.99 per hour, that is the wage earned by the most people.

SAMPLE TEST QUESTIONS (For answers, see pages 195-196.)

Multiple-Choice (Circle the correct answer.)

Use the line graph to answer questions 1 and 2.

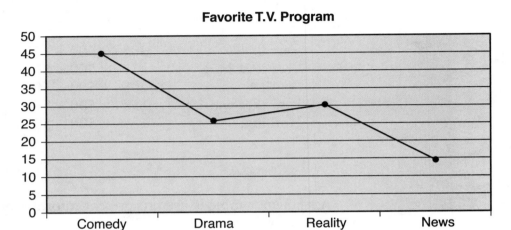

Favorite T.V. Program

1. Which type of television program was favored by slightly more than twice as much as news programs?

    A. Comedy

    B. Drama

    C. Reality

    D. News

2. Which type of program was chosen by one and a half times as much as reality programs?

    A. Comedy

    B. Drama

    C. Reality

    D. News

Use the circle graph to answer questions 3 and 4.

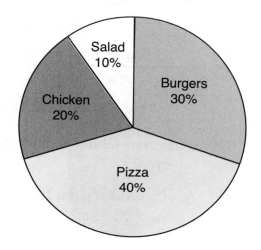

Favorite Cafeteria Food
150 Students Surveyed

3. How many students surveyed prefer pizza?

    A. 40

    B. 45

    C. 60

    D. 80

4. There are 600 students in the middle school. If the results remain the same throughout the entire middle school population, how many students would prefer burgers?

    A. 300

    B. 180

    C. 90

    D. 60

Use the frequency table to answer questions 5 and 6.

Favorite After School Activity	Frequency
Watch television	5
Play sports	7
Watch a movie or DVD	3
Play video games	6
Read a book or magazine	4

5. What after school activity was selected by 20% of those surveyed?

    A. Play video games

    B. Watch television

    C. Read a book or magazine

    D. Watch a movie or DVD

6. Which two activities equal 40% of those surveyed?

    A. Play video games and read a book or magazine

    B. Watch television and read a book or bagazine

    C. Play video games and watch a movie or DVD

    D. Watch television and play sports

Use the Birthday Month graph to answer questions 7 and 8.

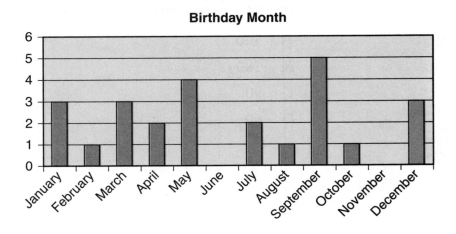

Birthday Month

7. How many people are surveyed?

   A. 25

   B. 24

   C. 20

   D. 5

8. Which statement is NOT true about the graph?

   A. The most popular birthday month is September.

   B. The total number of people with birthdays in June and November is 0.

   C. There are three times as many birthdays in April as there are in February.

   D. The same amount of people have birthdays in December and January.

Extended Constructed Response (Use the space provided to show your work.)

Construct a graph displaying the results from a survey asking 300 people their favorite ride at the amusement park. Make sure to title your graph, label your axes, and use an appropriate scale.

Roller coaster	45%
Merry-go-round	15%
Ferris wheel	10%
Water slide	30%

LESSON 3: DETERMINING THE PROBABLITY OF AN EVENT

Have you ever played a game that had some element of chance built into it? Think of some popular board games that you have played such as Monopoly, Risk, or Trouble. If you have ever needed to roll a certain number to win the game and rolled that die, did you think the number you needed would appear? I'm willing to say that sometimes you were lucky and other times you were not. That's what makes playing board games so much fun! It all involves some element of chance.

Probability is defined as the chance or likelihood that something particular will occur. Probability can be expressed as a ratio that defines a relationship between what you want to occur compared with how many possible outcomes there are.

In math, we need to determine the probability of an event to determine whether there is a good chance that a particular outcome will happen. In general, we can say that there is a range of possible outcomes for any particular event.

V.I.P. (Very Important Point)

A probability event is any experiment, game, etc. that involves chance. For example, an event could be rolling a die, picking a card, spinning a spinner, etc.

Impossible	Equally Likely	Certain
0%	50%	100%

If an event can never happen, we can say that it is impossible or has a 0% chance of occurring. If an event has a 50% chance of happening, then it is equally likely. If an event is absolutely going to happen, then it has a 100% chance of occurring.

Example 1

A standard number cube or die has six sides, and each side displays a number ranging from 1 to 6. If you roll the die, what is the likelihood that the number appearing on the die will be even?

Solution: Since there are six numbers on the die (1, 2, 3, 4, 5, 6) and three of them are even, there is a 50% chance (equally likely) that an even number will appear when the die is rolled.

Example 2

There are 20 people in your class. Every person writes his or her name on an index card and places it in a box. Your teacher will draw one name, and that person will win a free lunch. How would you determine the probability of winning the prize?

Solution: Since there are 20 students in your class, including yourself, then you would have a 1 in 20 chance (5%) of winning the prize. Remember, you want your name to be selected out of everyone in your class.

Example 3

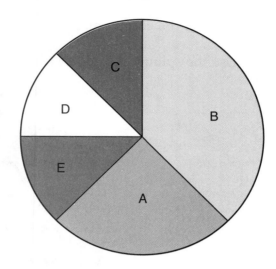

Examine this spinner. (Circle with 8 equal wedges = A, A, B, B, B, C, D, E)

V.I.P. (Very Important Point)

Theoretical probability is the chance that something should occur. For example, if you were to toss a coin ten (10) times, you should get five (5) heads and five (5) tails in theory. That is what *should* happen.

Experimental probability shows the results when conducting a probability experiment. For example, if you tossed a coin ten (10) times, how many heads and how many tails would appear? Sometimes the experimental probability equals the theoretical probability but often times it is slightly different. Experimental probability is what *does* happen.

What is the likelihood that the letter A or B will appear when the wheel is spun?

Solution: There are 8 equal spaces on the wheel. There are 5 spaces that contain an A or a B. The probability of getting an A or B is $\frac{5}{8}$.

Example 4

Refer to the spinner in Example 3. What is the likelihood that you would spin a B first followed by a D on the second spin?

Solution: When calculating compound probabilities, meaning that you want to determine the likelihood of something occurring first and then a second outcome appears, you must determine each individual probability and then multiply them together. The probability of getting a B is $\frac{3}{8}$ and the probability of getting a D is $\frac{1}{8}$.

Therefore, $\frac{3}{8} \cdot \frac{1}{8} = \frac{3}{64}$.

SAMPLE TEST QUESTIONS (For answers, see page 196.)

Multiple-Choice (Circle the correct answer.)

Use the spinner to answer questions 1 and 2.

Name Spinner

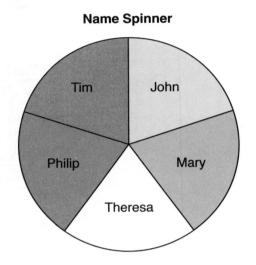

1. What is the probability that the spinner will land on a girl?

    A. $\dfrac{1}{5}$

    B. $\dfrac{2}{5}$

    C. $\dfrac{3}{5}$

    D. 1

2. What is the probability that the spinner will land on a girl first and then, when spun again, land on a boy?

    A. $\dfrac{6}{25}$

    B. 1

    C. $\dfrac{1}{2}$

    D. $\dfrac{3}{5}$

3. How many possibilities are there when you are rolling two number cubes and each cube is numbered with the digits 1–6?

   A. 6

   B. 12

   C. 36

   D. 216

4. What is the probability of rolling a number cube and having a 5 appear and then tossing a coin and having tails appear?

   A. $\dfrac{1}{12}$

   B. $\dfrac{1}{8}$

   C. $\dfrac{1}{6}$

   D. $\dfrac{1}{2}$

5. There are 15 marbles in a bag. Three marbles are red, two marbles are blue, four marbles are green, and six marbles are yellow. What is the probability of choosing a blue or green marble?

   A. $\dfrac{1}{15}$

   B. 0.2

   C. $\dfrac{4}{15}$

   D. 40%

6. Refer to question 5. Which outcome has the same probability as $\dfrac{2}{3}$?

   A. Choosing a yellow marble

   B. Choosing a yellow or green marble

   C. Choosing a red, blue, or green marble

   D. Not choosing a yellow or green marble

Extended Constructed Response (Use the space provided to show your work.)

You toss a single coin four times. What is the theoretical probability of tossing that coin and how does that compare with its experimental probability?

LESSON 4: PRINCIPLES OF COUNTING (COMBINATIONS AND PERMUTATIONS)

The fundamental counting principle states that you can determine the total number of possibilities for different events by multiplying the amounts of each event together.

Your local soccer team is getting ready for the upcoming season, and the coach needs to order uniforms for her players. She will order these uniforms from a local sporting goods store in town. There are 15 players on the team and four sizes available (Small, Medium, Large, and Extra Large). How many different uniforms are possible?

To calculate the total number of uniforms the store must have you must multiply the number of players by the different sizes of uniforms.

15	×	4	=	60
Number of Players		Sizes of Uniforms		Total Number of Uniforms

Permutations and combinations calculate the number of outcomes for a particular event. What makes them different is whether or not order is important. If order is important for an arrangement, you will calculate that as a permutation. However, if order is not important, then you will calculate it as a combination.

When might order be important? Order is important if you need to unlock a combination, create a password, or place a group of people in a particular order. Order isn't important if you are simply calculating how many items or arrangements there are. The main thing to remember is that if order is important, you must calculate that as a permutation.

There are special formulas or notations used to calculate combinations and permutations. It is important to understand the formulas for each and to be able to understand the keystrokes used to enter the formula on your calculator.

A permutation is calculated as $_nP_r$ where n represents the amount of items and r represents how many are chosen.

A combination is calculated as $_nC_r$ where n represents the amount of items and r represents how many are chosen.

Let's try a few examples and see how order can play a part in calculating these outcomes.

Example 1

A locker combination uses a dial with numbers ranging from 0 to 39 to create a three-number combination. When entered correctly, the lock will open. While any number can be used to create the first, second, and third numbers, a number can only be used once. How many different combinations can be created for this lock?

Solution: We can solve this problem by understanding that any one of the 40 numbers can be chosen for the first number. (Remember that even though the dial has numbers from 0 to 39, there are a total of 40 numbers.) Once that number is selected, there are 39 numbers remaining to choose from for the second number, and finally, there will be 38 numbers to choose from for the final number. Using the fundamental counting principle, we can simply multiply these numbers together.

$$40 \quad \times \quad 39 \quad \times \quad 38 \quad = 59{,}280$$

First number   Second number   Third number = Total number of
  possibilities   possibilities   possibilities   outcomes

Also, this problem can be solved using the permutation formula: $_{40}P_3 = 59{,}280$.

Example 2

Marta is creating a six key password that must consist of two letters followed by four digits. Letters and numbers can be repeated in the password. How many different combinations can be created using letters and numbers for this password?

Solution: The password must be created as follows:

Letter–Letter–Digit–Digit–Digit–Digit

There are 26 letters in the alphabet and 10 digits. Since letters and digits can be repeated, we can use the fundamental counting principle to multiply all of the outcomes.

$$26 \times 26 \times 10 \times 10 \times 10 \times 10 = 6{,}760{,}000$$
different passwords can be created

Example 3

Your teacher needs to select 3 people from your class of 10 to participate on the Student Council. How many different combinations of students can she select?

Solution: Since the order of selection is not important, we can use the combination formula to determine the total number of possibilities.

$$_nC_r = {_{10}C_3} = 120$$

V.I.P. (Very Important Point)

One thing to remember about using combinations is that you don't want to include all the "repeated" outcomes. For example, if your teacher chooses students A, B, and C, that would be the same as choosing students B, C, and A. It is as if we are saying that the "groups" are the same because order does not matter.

SAMPLE TEST QUESTIONS (For answers, see page 197.)

Multiple-Choice (Circle the correct answer.)

1. There are three boxes sitting on a table labeled *A*, *B*, and *C*. How many different ways can the boxes be arranged on the table?

    A. 27

    B. 9

    C. 6

    D. 1

2. During your class field day, 12 students compete in a quarter-mile race. How many different ways can people finish the race in the first, second, and third positions?

    A. 1,728

    B. 1,320

    C. 220

    D. 3

3. There are eight crayons in a box. If you need to use two of them, how many different sets of crayons can you select?

   A. 64

   B. 56

   C. 15

   D. 8

4. Mike, Sean, Travis, Jamal, and Jose all want to be captain of the school's soccer team. How many different ways can two co-captains be chosen?

   A. 10

   B. 20

   C. 25

   D. 120

5. You have to create a computer password using letters and numbers. The password must contain two letters and two numbers. Any letter or number can be used and repeated. How many different passwords can you create?

   A. 72

   B. 520

   C. 6,770

   D. 67,600

Short Constructed Response

6. Your teacher needs to arrange 4 students in a row. How many different combinations can your teacher create with these 4 students? _____

Extended Constructed Response (Use the space provided to show your work.)

Suzanne has three shirts (blue, red, and green), one pair of pants, one pair of jeans, one pair of shoes, and one pair of sneakers. Create a list or diagram showing how many different outfits she can create from the clothing described.

LESSON 5: VERTEX-EDGE GRAPHS AND VENN DIAGRAMS

Vertex-edge graphs are designed to show connections between a series of points or locations. By definition, a vertex is a point that connects two or more line segments. An edge is defined as the line in between two vertices.

In this section, you will see a series of vertex-edge graphs and be asked a series of questions pertaining to these graphs.

Example 1

Examine this vertex-edge graph.

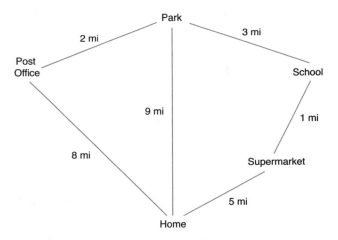

What is the shortest distance to get from home to the park? How can you go from home to school without passing the supermarket?

Solution: Before you answer the questions, take a moment to review the facts in this problem. Look at the vertices (buildings) as well as the distance between them (edges).

The shortest distance from home to the park is via a straight path, which is 9 miles. You can go from home to school without passing the supermarket by two different ways. First, you can go from home to the post office to the park and then to school (8 + 2 + 3 = 13 miles) or you can go from home to the park and then to the school (9 + 3 = 12 miles).

Example 2

The following vertex-edge graph is a model of a computer network for a local corporation. The computer at workstation A is connected to the other computers within the group. How many computers can workstation E communicate with other than workstation A?

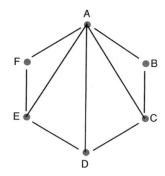

Solution: Reviewing the vertex-edge graph, we can see that workstation E is connected to workstations D and F along with A. However, the question specifically asks for the number of workstations other than A. Therefore, the correct answer is 2.

Another type of graph which may appear on the NJ ASK7 is called a Venn diagram. A Venn diagram consists of two or more circles, mostly interlocking circles, which show data pertaining to two independent items. Examine the Venn diagram below, review the questions, and follow along with the solutions.

Example 3

The following Venn diagram shows how many students in a particular seventh grade classroom own a dog and/or a cat.

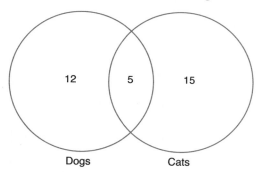

V.I.P. (Very Important Point)

When answering the last question, remember that the 5 students in the middle are students who have both a dog and a cat. A common mistake is to add that number twice. Remember that number represents the students who have a dog, a cat, or both. Nowhere in the problem did it show any information for students who do not own a dog or a cat.

■ How many students have a dog?
■ How many students have a cat?
■ How many students own both a dog and a cat?
■ How many students were surveyed all together?

Solution: In order to answer the questions, it may help you to write out the facts next to the diagram or on a separate sheet of paper.

■ In the circle labeled dogs, there are a total of 17 students (12 students in one part of the circle and 5 in the shared part).
■ There are 20 students who have a cat (15 students in one part of the circle and 5 in the shared part).
■ There are 5 students who have both a dog and a cat. That is shown by the connected or shared part of the diagram in the middle.

To calculate the total number of students surveyed, simply add up the numbers. 12 + 5 + 15 = 32 students.

SAMPLE TEST QUESTIONS (For answers, see pages 197–198.)

Multiple-Choice (Circle the correct answer.)

Use the diagram to answer questions 1 and 2.

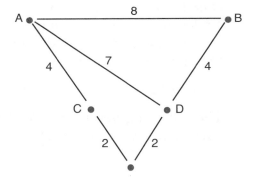

1. What is the shortest distance to get from vertex *B* to vertex *C*?

   A. 4
   B. 7
   C. 8
   D. 12

2. How far is it to get from *A* to *D* to *E*?

    A. 6

    B. 7

    C. 8

    D. 9

3.

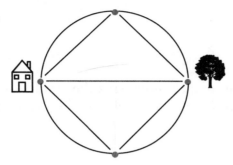

How many paths can be used to get from home to the park?

    A. 8

    B. 7

    C. 6

    D. 3

4.

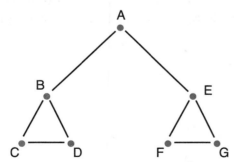

Point *A* represents the computer network administrator. This computer can communicate with all the computers in the network. Which computer station can ONLY communicate with two other computers?

    A. *B*

    B. *C*

    C. *E*

    D. All the computers can communicate with each other.

5.

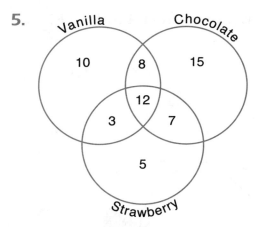

This Venn diagram displays the results of a school survey about ice cream. How many students like vanilla ice cream?

A. 13

B. 21

C. 33

D. 60

Extended Constructed Response (Use the space provided to show your work.)

You survey your seventh grade classmates and learn that 18 of them like baseball and 11 of them like football. However, 6 of your classmates like both baseball and football. Draw a Venn diagram displaying these results and determine how many students were surveyed in all.

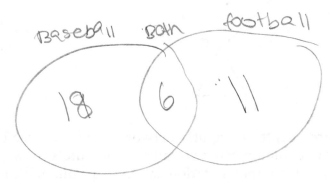

MATHEMATICAL PROCESSES

In this chapter, we will focus on various problem-solving strategies that can help you become more successful on the NJ ASK7. The mathematical processes described here are designed to give you a method for solving certain problems. The strategies used here are not the only ways to solve the problem but merely a different way to obtain a correct answer. Rather than providing sample test questions in this chapter, I will give you a problem set at the end of each lesson that will allow you an opportunity to practice the skills described.

LESSON 1: GUESS AND CHECK

Please understand that using guess and check to solve a problem is perfectly acceptable. In fact, guess and check is often used to solve problems where you must find two answers within a problem. It may take you a little more time in the long run to get the right answer, but the main point is that you will eventually get it. When using the guess and check method, you simply guess at an answer and check to see if it is correct. Using your estimation skills to help you make an educated guess will greatly reduce the amount of time you spend on a particular problem. Let's take a look at a few examples where you could use guess and check.

Example 1

A family has dogs and birds. There are 12 heads and 38 legs. How many dogs and birds does the family own?

Solution: You can solve this problem by using some logical reasoning. Since there are 12 heads, a good guess might be that there are 6 dogs and 6 birds.

6 dogs (24 legs) + 6 birds (12 legs) = 36 legs

Our guess is only two away from the actual answer. In order to get the right answer, you will need to have 2 more legs. Since a dog has two more legs than a bird, add one dog and take away one bird.

7 dogs (28 legs) + 5 birds (10 legs) = 38 legs

Example 2

Digging in between the seat cushions of the living room couch, your mother finds $1.35. She tells you that she only has quarters and nickels and has a total of 15 coins. She will give you the money if you can tell her how many of each coin she has. How many of each coin does she have?

Solution: We can create a table and use some guessing and checking to help us solve this problem.

Guess	Quarters ($.25)	Nickels ($.05)	Total Money	Total Coins
1	4 ($1.00)	7 ($.35)	$1.35	11
2	5 ($1.25)	2 ($.10)	$1.35	7
3	3 ($.75)	12 ($.60)	$1.35	15

There are 3 quarters and 12 nickels for a total of 15 coins equaling $1.35.

PROBLEM SET (For answers, see page 198.)

1. Mr. MacDonald's farm has horses and chickens. There are 20 heads and 64 legs. How many of each animal does Mr. MacDonald have?

2. In your piggy bank, you find quarters and dimes. There are 42 coins for a total of $8.10. How many of each coin do you have?

3. Your grandfather gives you a bag of quarters, nickels, and dimes. He tells you that the bag is worth $3.70. There are twice as many quarters as dimes. There is 1 less nickel than three times the number of dimes. How many of each coin is in the bag?

4. Mildred has cats and birds. There are 17 animals with a total of 44 legs. How many of each kind of animal does Mildred have?

LESSON 2: MAKING A LIST, DRAWING A DIAGRAM, WORKING BACKWARDS

Making a list is a useful problem-solving strategy to see all the possible answers to a given problem or to logically follow your steps as you solve the problem. One of the hardest parts of this strategy is to write everything down on your paper. Because the problems on NJ ASK7 are tricky at times, it is advised that you use your test booklet and/or your scrap paper to write out your list to help you get the right answer. Let's take a look at a few examples!

Example 1

Angelique is having some friends over for dinner. She wants to serve dinner at 7:15 P.M. She needs $\frac{3}{4}$ of an hour to cook the dinner. It will take her 10 minutes to set the table. She needs 35 minutes to get herself ready, and she needs a half hour to prepare the dinner. What is the latest she can start in order to serve dinner on time?

Solution: We can work backwards and make a list of the times and activities in order to determine when is the latest she can start.

Activity	Time	Serve Dinner
Serve dinner		7:15 P.M.
Cook dinner	45 minutes	6:30 P.M. (7:15 − 0:45)
Set the table	10 minutes	6:20 P.M. (6:30 − 0:10)
Get ready	35 minutes	5:45 P.M. (6:20 − 0:35)
Prepare the dinner	30 minutes	5:15 P.M. (5:45 − 0:30)

Example 2

Mark is getting ready for the seventh grade social, and he is trying to figure out what to wear. He has narrowed his choices to a blue shirt, a red shirt, and a green shirt. He could wear jeans or khakis and he could wear sneakers or dress shoes. Create a diagram or list to show how many different outfits Mark could create.

Solution: Create a list showing the different clothing options.

Shirts	Pants	Shoes
Blue	Jeans	Sneakers
Red	Khakis	Dress shoes
Green		

Based on the fundamental counting principle we learned earlier, we could multiply the amounts of shirts, pants, and shoes to get the total number of outfits. However, the problem specifically states that we must show a list in order to answer the question. Our list could look something like this:

1. Blue shirt–Jeans–Sneakers
2. Blue shirt–Jeans–Dress shoes
3. Blue shirt–Khakis–Sneakers
4. Blue shirt–Khakis–Dress shoes
5. Red shirt–Jeans–Sneakers
6. Red shirt–Jeans–Dress shoes
7. Red shirt–Khakis–Sneakers
8. Red shirt–Khakis–Dress shoes
9. Green shirt–Jeans–Sneakers
10. Green shirt–Jeans–Dress shoes
11. Green shirt–Khakis–Sneakers
12. Green shirt–Khakis–Dress shoes

PROBLEM SET (For answers, see page 198.)

1. Karen returned home after a day of shopping. She bought a new sweatshirt for $20.00, a pair of sneakers for $45.00, and a new CD for $15.00. She also bought lunch for $8.00. If she came home with $12.00, how much did she have at the beginning of the day?

2. Corey and Dustin go to a baseball game. They bring a bag of peanuts with them. Corey ate one-half of the peanuts and Dustin ate three-fifths of what was left. If there are 16 peanuts in the bag after Corey and Dustin have eaten, how many peanuts were in the bag at the beginning?

3. Tyrell came home at 7:00 P.M. He played basketball with his friends for 45 minutes. He studied at the library for an hour and a half. He also watched television at a friend's house for a third of an hour. What time did he get to his friend's house to watch television?

4. Your class has won a school spirit contest! For winning, your class is awarded an ice cream sundae party. There are three flavors of ice cream, two sauces, and four toppings to choose. You can only select one flavor of ice cream, one sauce, and one

topping. How many different combinations of sundaes can you make?

LESSON 3: DETERMINING THE REASONABLENESS OF AN ANSWER

You should always check your work. I know there were times when I was in school when the teacher would call on me, and in a rush to give the right answer, I blurted out something that didn't completely make sense! It's okay to be either quick or slow with your answers, but most importantly, your answers have to be reasonable.

NJ ASK7 creates questions that test your ability to make certain that an answer is reasonable. These questions can be either multiple-choice or constructed response, and they can be rather tricky. Your goal is to always check your work and make certain that your answers are reasonable. Let's take a look at some examples that will make this concept clearer for you.

Example 1

Jose is checking Miguel's work on a math quiz. One of the problems on the test reads:

What is the sum of 149 + 289 + 73 + 581?

A. The sum is between 500 and 1,000.

B. The sum is between 1,000 and 1,500

C. The sum is between 1,500 and 2,000.

D. The sum is between 2,000 and 3,000.

Miguel chose answer C. What should Jose tell Miguel about his answer choice and why.

Solution: Jose should tell Miguel that answer C is incorrect. The correct answer is B because the sum equals 1,092 and that is between 1,000 and 1,500. In addition, Jose should tell Miguel that it appears he made a computation or keystroking mistake when using the calculator. It looks like Miguel thought all of the numbers were three digits and thus, thought 73 was actually 730.

Example 2

Miguel is checking Jose's work on a math quiz. Jose's answer to the problem, "Find the perimeter of a rectangular region that has a width of 12 units and a length of 15 units" is 180 units. What should Miguel tell Jose?

Solution: Miguel should tell Jose that his answer is incorrect because perimeter measures the distance around a figure. If a rectangular region has a width of 12 units and a length of 15 units, its perimeter couldn't be 180 units because the dimensions are too small to have a perimeter of 180 units. Miguel should tell Jose that he calculated the region's area instead of its perimeter.

PROBLEM SET (For answers, see page 198.)

1. What is a reasonable estimate for 32×19?

    A. 55

    B. 300

    C. 600

    D. 900

2. Mark has 47 quarters, and Carolyn has 55 quarters. Carolyn has how much more money than Mark?

    A. $8.00

    B. $2.00

    C. $.25

    D. $.05

3. Pierre bought 3 shirts at $18.99 each, 2 ties for $11.99 each, and a pair of shoes for $79.99. What is a reasonable estimate of how much he paid for his clothes?

    A. $160

    B. $110

    C $100

    D $85

4. Every person in Ms. Stewart's class gets three cookies. If there are twenty-four students in her class, how many cookies are given in total?

   A. 24

   B. 27

   C. 72

   D 96

LESSON 4: HOW TO SOLVE A WORD PROBLEM

Word problems are difficult. Not only do you have to be able to do the math but also you must extract the key components of the problem in order to solve what is being asked. For some, the difficulty lies in being able to determine what exactly is being asked and then to do the math correctly! I want to give you a few suggestions that can help you solve word problems. You can use these strategies for success on the NJ ASK7 but also in your math classes.

It should come as no surprise, but if you don't understand the words in the problem, it will be difficult to solve. Even though you can't reach for a dictionary or ask for the meaning of a particular word, you can use the context clues (words around the word you don't know) to give you an idea. Some people find it useful to underline the question so it visually stands out from the problem and will always remind you of what is being asked. Once you have read the question and understood its meaning, it's time to get to work and find that answer.

There are four key steps to solving a problem. You may have seen something similar to this in your math classes because the process is uniformly known.

Step 1: Understand the Facts

- Make sure you have all the facts to solve the problem.
- Cross out any useless or unnecessary information in the problem.
- Identify the question.

Step 2: Determine Your Plan for Solving the Problem

- Now that you have all of the facts, what problem-solving strategy will you implement to solve the problem?

Step 3: Do the Work

▪ Put your plan to work. Do the math and get an answer.

Step 4: Check Your Work

▪ Make sure your answer is reasonable and you have done the math correctly.
▪ If this is a multiple-choice question, make certain that you have correctly colored in the right answer on your answer sheet.
▪ If this is a constructed response question, make certain that you have answered all the components of the question and have written your response neatly.

Let's take a look at a word problem and use the method described above to solve it.

Example

Josephine stops at the gas station to fill up her car. It has been 384 miles since the last time she put gas in her car. The car's gas tank holds 14.6 gallons. Each gallon of gas costs $2.89. How many miles does Josephine's car get per gallon? (Round your answer to the nearest tenth.)

Solution:

Step 1: Understand the Facts

▪ It has been 384 miles since the last time Josephine filled up her gas tank.
▪ It takes 18.6 gallons to fill up the tank.
▪ Don't need to know the price per gallon.
▪ How many miles per gallon?

Step 2: Determine Your Plan for Solving the Problem

▪ Divide the number of miles driven by the total number of gallons.

Step 3: Do the Work

▪ $384 \div 18.6 = 20.645161$
▪ The question states to round the answer to the nearest tenth. Therefore, the answer is 20.6.

Step 4: Check Your Work

■ Since division is used in this problem to get the answer, we can check by multiplying: $20.6 \times 18.6 = 383.16$.

■ Since the product is close to the number of miles driven, the answer seems reasonable.

PROBLEM SET (For answers, see page 199.)

Directions: Use the steps described to help you solve the following word problems.

1. Cathy loves to ride her bicycle for a good workout. She estimates she can burn 400–600 calories an hour. During any week she rides between 8 and 12 hours. How many calories did she approximately burn riding her bike during the week?

2. You are responsible for making house numbers that appear on the front door. You will make numbers for houses ranging from 100 to 200. How many zeros will you need to make all the house numbers?

3. If you roll two number cubes simultaneously and each number cube has the digits 1–6 on each, how many different ways can a 2 appear on one or both cubes?

4. A backyard measures 25 feet by 18 feet. If a 3-foot walkway is created around the perimeter of the yard, what is the area of the backyard that can be used to plant grass seed?

LESSON 5: HOW TO ANSWER AN EXTENDED-REPSONSE QUESTION

Answering extended-response questions on the NJ ASK7 is very important. One thing to keep in mind is that these questions are worth more points than a multiple-choice question. Additionally these questions are scored based on a rubric that is designed to award partial credit. It is extremely important to make your best attempt at these questions and write anything you feel is correct. It is far better to get some credit for your answer than nothing. And please remember this. NEVER, NEVER, NEVER leave an extended-response question blank.

Below are some guidelines from the New Jersey Department of Education's website (http://www.state.nj.us/education/assessment/ms /5–7/specs/math7.htm, accessed September 17, 2007) regarding the development of scoring rubrics for the extended-response questions.

RULES FOR DEVELOPING SCORING RUBRICS

- An item-specific scoring rubric will be used for each extended-response item.
- Information from student sample papers will be used to refine these scoring rubrics.
- Scoring rubrics will follow a "focused holistic" model in which the score for the response is based on overall quality in relation to the strand; individual score point elements must arise from the specific individual requirements of the item and its target.
- Scoring rubrics for content strand items will focus on conceptual understanding, application of appropriate procedures, and accuracy.
- Scoring rubrics for process items will focus on such aspects as clarity, effectiveness, reasonableness, selection of useful procedures, and/or degree to which solutions are viable.
- Extended-response items will be scored with a four-level scoring rubric (0–3).

Here are a few suggestions that can help you to answer the constructed response questions.

- Take the question and turn it into a topic sentence. With questions that require a written response, it looks good to begin your response with a complete sentence that restates the question.
- If a question requires you to show work or to show your calculations, be very specific and careful as you write your calculations.
- When a question states to write a list, draw a diagram, make a table, etc., be certain to provide enough information in the list, diagram, or table to answer the question completely.
- Always write your response neatly and clearly. If the person scoring your response cannot read what you have written, your response will be marked wrong.

ANSWER KEY TO PRACTICE QUESTIONS, CHAPTERS 1-5

CHAPTER 1

LESSON 1: INTEGERS

Multiple-Choice

1. B $(-7) + (-2) = (-9)$
2. C $(-3) + 5 = 2$
3. D $(-3) \cdot (-5) = 15$; $(-3) + (-5) = (-8)$
4. B $+5 + (-7) + 11 = +9$
5. C $\$250,000 + (-\$180,000) = +\$70,000$

Short Constructed Response

6. 1 $(-4) + 7 + (-8) + 11 + (-5) = 1$
7. $15.00 $28 - $4 - $9 = 15

Extended Constructed Response

The highest score is 9. The lowest score is (–9). It is possible to get zero. One possible way to score 0 is 0 + 1 + (–1). The highest possible score with a tic-tac-toe is vertically in the center to equal 5.

LESSON 2: FRACTIONS, DECIMALS, AND PERCENTS

Multiple-Choice

1. **C** Convert fraction to decimal. $15 \div 24 = 0.625$
2. **A** Martha has not read 80 pages. $\dfrac{80}{200} = \dfrac{2}{5}$
3. **C** There are 36 out of 64 boxes shaded. $\dfrac{36}{64} = \dfrac{x}{100}; x = 56.25\%$
4. **B** $57,545 \cdot 97\% = 55,818.65 \approx 55,819$
5. **B** Perform conversions to determine that they are all equal.
6. **D** $\$173,000,000 \cdot 5\% = \$8,650,000$
7. **C** There are 9 marbles that are either blue or green out of 16.
8. **A** She did not answer 15% of the questions correctly.
   3 out of 20 = 15

Short Constructed Response

9. $\dfrac{13}{40}$ $\qquad \dfrac{325}{1,000} = \dfrac{13}{40}$

10. 0.675 $\qquad 27 \div 40 = 0.675$

Extended Constructed Response

Answers may vary. Point A is located at 1.2, and Point B is located at 1.5.

LESSON 3: WHOLE NUMBERS AND EXPONENTS

Multiple-Choice

1. **A** $9 \times 9 \times 9 \times 9 = 9^4$
2. **A** $5^4 = 625; 5^2 = 25; 625 + 25 = 650$
3. **B** $12^3 = 12 \times 12 \times 12$
4. **C** $4^0 = 1$

5. C Day 1 = 3 pennies, Day 2 = 9 pennies, Day 3 = 27 pennies, Day 4 = 81 pennies, Day 5 = 243 pennies

6. B 10:00 A.M. = 500, 10:30 A.M. = 1,000, 11:00 A.M. = 2,000, 11:30 A.M. = 4,000, 12:00 noon = 8,000

7. B 8^5 = 32,768

8. B Evaluate each answer. 8^2 = 64

Short Constructed Response

9. 5 5^3 = 125

Extended Constructed Response

$$4^2 + 3^4 = 2^4 + 4^3$$
$$16 + 81 = 16 + 64$$
$$97 = 80$$

This answer is false because 97 does not equal 80. While addition is commutative, switching the base and exponent of a number does not always provide the same value. Therefore, Brad's answer is incorrect.

LESSON 4: COMPARING AND ORDERING INTEGERS

Multiple-Choice

1. C 5.5 feet = 66 inches, which is the shortest of all amounts.

2. A 0.28 = 28%, which is the greatest of all amounts.

3. C Rank in order from least to greatest and $3\frac{1}{5}$ is third.

4. B Point Y is less than 0 but greater than −1.

5. C When expressed in the same form, $\frac{9}{10}$ is the largest.

6. C $\frac{14}{15} > \frac{14}{16}$ because when the numerators are the same, the fraction with the smaller denominator is larger.

Short Constructed Response

7. Any value which is greater than 0.77 and less than 0.8

Extended Constructed Response

1. 25% of 60 is equal to 60% of 25. The problems equal 15 because both problems use multiplication and use the same numbers. Since multiplication is commutative, meaning that we can reverse the numbers and still get the same product, the answers are the same.

2. A number line must be drawn and correctly labeled to complete the question. Answers may vary.

LESSON 5: ORDER OF OPERATIONS

Multiple-Choice

1. **B** Follow the order of operations.

2. **A** Follow the order of operations.

3. **B** Perform multiplication before addition and subtraction.

4. **C** Working from left to right, you would perform multiplication second.

5. **D** Follow the order of operations.

6. **A** Perform what is inside of the parentheses first.

7. **A** Follow the order of operations.

Short Constructed Response

8. 88 $7^2 - 12 + 33 + 9 \times 2$

$49 - 12 + 33 + 9 \times 2$

$49 - 12 + 33 + 18$

$37 + 33 + 18$

$70 + 18$

88

Extended Constructed Response

The correct solution, line by line, is

$$13 - 9 + 18 \times 3 - 7^0 \times 8$$
$$13 - 9 + 18 \times 3 - 1 \times 8$$
$$13 - 9 + 54 - 1 \times 8$$
$$13 - 9 + 54 - 8$$
$$4 + 54 - 8$$
$$58 - 8$$
$$50$$

LESSON 6: ABSOLUTE VALUE

Multiple-Choice

1. **C** 7 and (–7) are 7 units away from zero.
2. **A** The absolute value of –14 is 14, so 14 – 5 = 9.
3. **B** $\left|7-4\right| = 3$ and $\left|-1+4\right| = 3$

4. **A** $\left|2^0 + 1^3\right| = \left|1+1\right| = 2$
5. **D** The absolute value of any number is always positive.
6. **D** There are two numbers that are 7 units from 2. They are –5 and 9.

Short Constructed Response

7. –4
$$\left|-5\right| - x = 9$$
$$5 - x = 9$$
$$-x = 4$$
$$x = -4$$

Extended Constructed Response

By definition, the absolute value of a number represents how many units the number is from zero. Both +2 and (−2) have an absolute value of two as they are both located two units from zero on the number line.

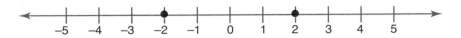

LESSON 7: RATIO AND PROPORTION

Multiple-Choice

1. **A** Make a proportion relating the $\dfrac{\text{part}}{\text{whole}}$.

2. **B** $\dfrac{5}{8} = \dfrac{x}{32}$; $x = 20$

3. **A** $\dfrac{2}{18} = \dfrac{x}{54}$; $x = 6$

4. **C** $\dfrac{3}{7} = \dfrac{x}{112}$; $x = 48$

5. **B** The new perimeter is 42 inches. $42 \div 3 = 14$

6. **A** $\dfrac{5}{12} = \dfrac{15}{x}$; $x = 36$

7. **C** Multiply the original amount of raisins by 4.

Short Constructed Response

8. 20
$$\dfrac{5}{8} = \dfrac{12.5}{x}$$
$$5x = 100$$
$$x = 20$$

LESSON 8: RATIONAL NUMBERS

Multiple-Choice

1. **B** Choices A, C, and D are all between 1.0 and 1.1.

2. **A** Choices B, C, and D are not between 2.3 and 2.4.

3. **C** Choices A, B, and D are not between $\frac{3}{4}$.

4. **D** Convert each number to its decimals equivalent.

5. **A** Choices B, C, and D are not between $\frac{8}{10}$ and $\frac{9}{10}$.

6. **C** $\frac{3}{8} = 0.375$

7. **B** 2.12 is the only choice less than $2\frac{1}{8}$.

Short Constructed Response

8. Any number greater than 1.7 and less than 1.8. (A possible answer would be 1.75.)

Extended Constructed Response

Answers may vary. Check for the appropriateness of the number line. Possible correct numbers are: 0.35, 0.38, $\frac{7}{20}, \frac{38}{100}$. Note that a correct response will include two different numbers.

LESSON 9: ESTIMATION AND ROUNDING

Multiple-Choice

1. **B** Since the digit in the tens place is greater than 5, increase the digit in the hundreds place by 1.

2. **C** $70 \times 20 = 1,400$

3. **C** 35,000 and 44,999 each round to 40,000.

4. **B** $70 + $40 + $60 = $170, which is closer to $150.

5. A Round each number down so $\$200 \div 20$.

    $\dfrac{1}{4} = \dfrac{x}{13}$; $x = 3.25$ inches

6. B $20{,}000 + 60{,}000 + 10{,}000 = 90{,}000$

7. C $50 \times 20 = 1{,}000$

8. B Since the digit in the hundreds place is greater than 5, increase the digit in the thousands place by 1.

Short Constructed Response

9. 1.0 Since the digit in the thousandths place is 5 or greater, increase the digit in the hundredths place by 1.

Extended Constructed Response

The actual price range for this new car is between $\$21{,}450$ and $\$21{,}549$ because all numbers in this range round to $\$21{,}500$.

LESSON 10: PERCENT OF DISCOUNT

Multiple-Choice

1. C $\dfrac{21}{60} = \dfrac{x}{100}$; $x = 35$

2. B $\$15 \cdot 20\% = \$3$; $\$15 - \$3 = \$12$

3. D $\$3{,}200 \cdot 6\% = \$180$ and will not yield the necessary $\$200$ so the best answer choice is D.

4. A $\$105 \cdot 70\% = \$73.50$

5. D $\dfrac{\$270.00}{x} = \dfrac{80}{100}$; $x = \$337.50$

6. A $\dfrac{\$140}{\$700} = \dfrac{x}{100}$; $x = 20$

7. C $\dfrac{x}{200} = \dfrac{70}{100}$; $x = \$140$

Short Constructed Response

8. 30% $\dfrac{\$75}{\$250} = \dfrac{x}{100}$; $x = 30\%$ Place $75 in the proportion because it represents the amount of the discount.

Extended Constructed Response

Jameka and Caitlyn paid the same price for their necklaces.

Jameka's Necklace: $60 × 40% = $24 → $60 − $24 = $36

Caitlyn's Necklace: $80 × 55% = $44 → $80 − $44 = $36

LESSON 11: SQUARE NUMBERS AND CUBIC NUMBERS

Multiple-Choice

1. D $8^2 = 64$ and 64 is greater than 50.

2. B $7 \cdot 7 \cdot 7 = 343$

3. B Multiplying three negative numbers will yield a negative product.

4. B $5 \cdot 5 \cdot 5 = 125$

Short Constructed Response

5. 11 $11 \cdot 11 \cdot 11 = 1{,}331$

Extended Constructed Response

Square numbers greater than 0 and less than 100 are 1, 4, 9, 16, 25, 36, 49, 64, and 81.

Cubic numbers greater than 0 and less than 100 are: 1, 8, 27, and 64.

The numbers 1 and 64 are both square numbers and cubic numbers that are greater than 0 and less than 100.

LESSON 12: DIVISIBILITY

Multiple-Choice

1. D Use divisibility rules.

2. C $(4 \cdot 5) + (3 \cdot 6) = 38$

3. A Use divisibility rules.

4. C $197 \div 7 = 28r1$

5. C $176 \div 44 = 4$

6. A $150 \div 6 = 25$

7. B $(7 \cdot 2) + (6 \cdot 3) = 32$

Short Constructed Response

8. 21, 24, 30, 33, or 39   Numbers are divisible by 3 but not by 9.

Extended Constructed Response

The only way the teacher can divide the class of 23 students into groups is to leave the class as one large group. Since 23 is a prime number, it is only divisible by 1 and itself. Therefore, the teacher must leave the group as the entire class.

LESSON 13: PRIME AND COMPOSITE NUMBERS/PRIME FACTORIZATION

Multiple-Choice

1. C $30 = 2 \cdot 3 \cdot 5$

2. D $11 \cdot 11 = 121$; 121 is a prime number; therefore, it only has one prime factor.

3. B $2^4 \cdot 3 = 48$

4. B 2 and 13 are prime factors of 52.

5. D 79 is a prime number less than 80.

6. D 29 is a prime number greater than 25 and less than 30.

7. **A** 121 is a composite number. It is also a square number.

8. **B** 53 is the only prime number in the answer choices.

9. **D** 70 has three different prime factors: 2, 5, and 7.

Short Constructed Response

10. Answers may vary but possible answer choices are 30 or any multiple of 30.

Extended Constructed Response

$$18 = 5 + 13$$
$$28 = 5 + 23$$
$$40 = 3 + 37$$

CHAPTER 2

LESSON 1: TYPES OF POLYGONS

Multiple-Choice

1. **B** A quadrilateral has four sides.

2. **C** Squares and rectangles have right angles. Trapezoids have only one pair of parallel sides.

3. **A** Opposite angles in a parallelogram are congruent. Angle C is opposite Angle A.

4. **D** A trapezoid has only one pair of parallel sides.

5. **D** $180(10 - 2) = 1,440$

6. **A** A dodecagon has twelve sides.

Short Constructed Response

7. $105°$ $540 - (140 - 80 - 100) = 210$
            $210 \div 2 = 105$

Extended Constructed Response

The parallelogram needs 2 triangles.

The trapezoid needs 3 triangles.

The hexagon needs 6 triangles.

LESSON 2: LINES AND ANGLES

Multiple-Choice

1. **A** A 45° angle is an acute angle by definition.

2. **C** Two rays in opposite directions form a straight line.

3. **A** This is a line.

4. **C** $2x + 10 = 60$
$$2x = 50$$
$$x = 25$$

5. **C** $5n + 50 = 90$
$$5n = 40$$
$$n = 8$$

Extended Constructed Response

Answers may vary. Construct a 70° angle and label accordingly. Then, construct a 20° angle.

LESSON 3: SIMILAR AND CONGRUENT FIGURES

Multiple-Choice

1. **D** $15 \cdot 3 = 45$

2. **B** 6 in. × 8 in. is not similar to 4 in. × 6 in. because the dimensions are not increased by the same proportion.

3. **D** Perimeter is a line measure so it is increased by 2.

4. **C** The area of $\square LMNO = 16^2 = 256$ in.^{2} and the area of $\square EFGH = 8^2 = 64$ in.^{2}. $256 \div 64 = 4$

5. D All of the statements are true.

6. C Similar figures must have the same angle measure.

Short Constructed Response

7. 8 The length of each side is 16 inches and half of that is 8 inches.

LESSON 4: CIRCLES

Multiple-Choice

1. C $3 \cdot 3.14 \cdot 5 = 47.1$

2. A $3.14 \cdot 20^2 = 1{,}256 \text{ ft}^2$

3. C $(3.14 \cdot 12^2) - (3.14 \cdot 4.5^2) = 388.575 \text{ m}^2$

4. B $\sqrt{(380 \div 3.14)}$

5. C Substitute values for radii.

Short Constructed Response

6. 7 $\begin{aligned} 153.86 \div 3.14 &= 49 \\ \sqrt{49} &= 7 \end{aligned}$

LESSON 5: TRANSFORMATIONS: TRANSLATIONS, ROTATIONS, AND REFLECTIONS

Multiple-Choice

1. B Move points on the coordinate plane.

2. D Take the opposite of the x-coordinate.

3. A Turn figure $\frac{1}{4}$ clockwise.

4. C Move points on the coordinate plane.

5. B Answer is incorrect because the coordinates are reversed.

Extended Constructed Response

To transform point A to A', you can reflect the figure over the y-axis and then move the figure 6 units down. You could also reflect the figure over the x-axis and move the figure 6 units to the right.

LESSON 6: CUSTOMARY MEASUREMENT

Multiple-Choice

1. C 6 feet 2 inches − 5 feet 6 inches = 8 inches
2. C 128 · 16 = 2,048
3. B 4 hours 11 minutes − 1 hour 15 minutes = 2 hours 56 minutes
4. D 13,200 ÷ 5,280 = 2.5
5. A 12,000 ÷ 2,000 = 6

Short Constructed Response

6. $\frac{1}{2}$ mile $\frac{3}{4} + 1\frac{1}{8} + 1\frac{5}{8} + x = 4$

$$3\frac{1}{2} + x = 4$$

$$x = \frac{1}{2}$$

LESSON 7: METRIC MEASUREMENT

Multiple-Choice

1. C 95 · 1.6 = 152
2. D 14 · 2.2 = 30.8
3. B 21 · 1,000 = 21,000
4. A 0.16 kL is not equivalent to 16 L.
5. D This is the only true statement.

Short Constructed Response

6. 1.6   1,600 m = 1.6 km

Extended Constructed Response

■ It would take you and your family 47 hours to drive from Newark, New Jersey to Los Angeles, California.

■ It would take 557.75 hours to run from Newark, New Jersey to Los Angeles, California.

LESSON 8: AREA AND PERIMETER OF POLYGONS

Multiple-Choice

1. B $P = 2l + 2w$; $P = 2(18) + 2(15)$; $P = 66$

2. C Area = 80 m²; 25% of room is not covered. $80 \cdot 25\% = 20$

3. A 15 ft + 36 ft + 39 ft = 90 ft

4. B $A = \dfrac{1}{2}h(b_1 + b_2)$

   $A = \dfrac{1}{2}(14)(8 + 4)$

   $A = 84$ in.²

5. B $(22 \cdot 18) + (15 \cdot 18) + (33 \cdot 21) = 1{,}359$ ft²
   1,359 ft² ÷ 350 ft² ≈ 3.9 gal

Short Constructed Response

6. 4 72 ÷ 18 = 4

Extended Constructed Response

Jeri is correct because if you increase the length by 10% and decrease the width by 10%, you will not have the same area of the original rectangle. Jeri proves her answer by determining that the original area of the rectangle is 600 square feet. If you increase the length by 10%, the new length is 44 feet. If you decrease the width by 10%, the new width is 13.5. When you multiply 44 × 13.5, the new area is 594 square feet, which is less than the original area.

LESSON 9: AREA OF COMPOSITE FIGURES

Multiple-Choice

1. B $A = (12 \cdot 12) - (32 + 32) = 80$ units^2

2. A $A = 24^2 - 4(3.14 \cdot 6^2)$

3. C $A = (18 \cdot 6) - (3 \cdot 11)$

4. B $A = (22 \cdot 10) + (\frac{1}{2} \cdot 5 \cdot (22 + 12))$

Extended Constructed Response

In order to determine the area of the living room, I drew two lines across the top of the diagram to make the living room appear to be one large rectangle. I found the area of the large rectangle and then subtracted the area of the two smaller rectangles that were formed at the top.

The area of the large rectangle is $27 \times 15 = 405$ ft^2.

The area of the 2 small rectangles is $2 \times 3 \times 9 = 54$ ft^2.

Large rectangle – 2 Small rectangles = 405 ft^2 – 54 ft^2 = 351 ft^2

If carpeting sells for $4.50 per square foot, multiply the square footage of the living room by $4.50 to get the total cost of the carpet.

$$351 \times \$4.50 = \$1,579.50$$

Since the Ask family has a $1,500 budget for carpet, they do not have enough money to buy the new carpet.

LESSON 10: VOLUME OF PYRAMIDS, CONES, PRISMS, AND CYLINDERS

Multiple-Choice

1. C $V = \frac{1}{3}(18 \cdot 18 \cdot 36)$

2. C $V = \frac{1}{3}(3.14)(8^2)(12)$

3. D $V = 7^3$

4. **A** $V = (3.14)(2^2)(14)$

Short Constructed Response

5. 100.48 cm^3 $V = \dfrac{1}{3}(3.14)(4^2)(6)$

Extended Constructed Response

One possible answer is length = 4 units, width = 8 units, and height = 6 units.

LESSON 11: SURFACE AREA OF PRISMS AND CYLINDERS

Multiple-Choice

1. **D** $SA = (2 \cdot 4 \cdot 4) + (2 \cdot 4 \cdot 4) + (2 \cdot 4 \cdot 4)$

2. **B** $SA = (2 \cdot 12 \cdot 4) + (2 \cdot 4 \cdot 2) + (2 \cdot 12 \cdot 2)$

3. **B** $SA = (2 \cdot 3.14 \cdot 8 \cdot 15) + (2 \cdot 3.14 \cdot 8^2)$

4. **B** $SA = (2 \cdot 3.14 \cdot 3\frac{3}{4} \cdot 15) + (2 \cdot 3.14 \cdot (3\frac{3}{4})^2)$

5. **D** $SA = (2 \cdot 3 \cdot 6) + (2 \cdot 6 \cdot 9) + (2 \cdot 3 \cdot 9)$

Short Constructed Response

6. 288 cm^2 $SA = (2 \cdot 12 \cdot 6) + (2 \cdot 6 \cdot 4) + (2 \cdot 12 \cdot 4)$

7. 3,165.12 cm^2 $SA = (2 \cdot 3.14 \cdot 12 \cdot 30) + (2 \cdot 3.14 \cdot 12^2)$

Extended Constructed Response

Answers may vary. Simply reducing each dimension will yield a smaller surface area. Length = 7 inches, width = 5 inches, height = 3 inches.

$$SA = (2 \cdot 7 \cdot 5) + (2 \cdot 5 \cdot 3) + (2 \cdot 7 \cdot 3)$$

$$SA = 142 \text{ in.}^2$$

$$(2 \cdot 7 \cdot 5) + (2 \cdot 5 \cdot 3) + (2 \cdot 7 \cdot 3)$$

LESSON 12: SCALE DRAWINGS / DILATIONS

Multiple-Choice

1. B $(-1 \cdot 3, 2 \cdot 3)$
2. A Each side of the square measures 4 units. $4^2 = 16$ units^2
3. C $\dfrac{30}{75} = \dfrac{48}{x}$; $x = 120$ feet
4. D $5 \cdot 2 = 10$
5. C $(6 \cdot \dfrac{1}{2}, 0 \cdot \dfrac{1}{2})$

Short Constructed Response

6. 3.25 inches $\qquad \dfrac{1}{4} = \dfrac{x}{13}$; $x = 3.25$ inches

CHAPTER 3

LESSON 1: EVALUATING ALGEBRAIC EXPRESSIONS AND EQUATIONS

Multiple-Choice

1. C Evaluate the expression to equal 28.
2. B $a^3 = 64$; $4a^2 = 64$
3. B $3d^2 = 3 \cdot 5^2 = 75$
4. A Substitute information to create the formula.
5. A $C = \$.37x$; $C = \$.37(14) = \$5.18$
6. C Substitute information to create the expression.

Short Constructed Response

7. 96 $\quad 3a^2b = 3(4^2)(2) = 3(16)(2) = 96$

Extended Constructed Response

57 is not the correct solution to the expression. Paulina replaced the variable with the correct number but didn't multiply the coefficient with the substituted value. For example, she replaced the values for the variables in the expression $4r + 3s - 2t$ to make it equal $46 + 34 - 23 = 57$. The correct way to evaluate this expression is to multiply the coefficient with the numerical value of the variable.

LESSON 2: SOLVING ALGEBRAIC EQUATIONS

Multiple-Choice

1. B To isolate the variable, you must subtract 7 from both sides.
2. D $6(7 + 2) = 54$
3. B $\$96 \div \$12 = 8$
4. C Substitute the values into the equation.
5. B Miles $= (\$50.00 - \$22.50) \div \$.15 = 183.\overline{3}$
6. B $3(9 + 7) = 48$

Short Constructed Response

7. 21
$$4x + 17 = 101$$
$$4x = 84$$
$$x = 21$$

Extended Constructed Response

- ▪ $\$1.25X = \$20$
- ▪ You can download 16 songs.
- ▪ $\$1.25X = \$50$ You can download 40 songs.

LESSON 3: DETERMINING THE RULE FOR A PATTERN

Multiple-Choice

1. A Substitute values for n.
2. A Substitute values for n.
3. A Use the answer choices.

4. D Substitute values to get the correct expression.

5. D Make a diagram or chart.

6. C $12 \cdot 15 = 180$; $24 \cdot 30 = 720$

Short Constructed Response

7. 5 Make a diagram or chart.

Extended Constructed Response

Minutes	Cost
250	$30.00
300	$35.00
350	$40.00
400	$45.00

The formula for calculating the cost of the cell phone bill is

$$\text{Cost} = \$30.00 + \$.10(M - 250) \text{ where } M \geq 250$$

The alternate plan would be less expensive because it would cost $50.00 for 500 minutes as opposed to $55.00 for 500 minutes under the current plan.

LESSON 4: TYPES OF PATTERNS AND SEQUENCES

Multiple-Choice

1. B There are 12 complete sets (GYBR) and then 2 beads into the next set.

2. B Make a chart or table.

3. C $3,500 \div 2 = 1,750$
$1,750 \div 250 = 7$

4. C Year 1 = 5,000, Year 2 = 10,000, Year 3 = 20,000, Year 4 = 40,000

5. C Make a chart or table.

6. **C** There are 5 rows of 6 blocks in each row in the fifth arrangement.

7. **A** Each arrangement is a square number: $8^2 = 64$.

8. **D** Fibonacci Sequence: 2, 3, 5, 8, 13, 21, 34, . . .

Short Constructed Response

9. 128   1, 2, 4, 8, 16, 32, 64, 128, . . .

Extended Constructed Response

▪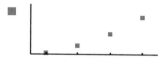

▪ (5, 25), (6, 36), (7, 49)

▪ The relationship is not linear because the points do not exist in a straight line.

LESSON 5: WRITING AN EXPRESSION OR EQUATION

Multiple-Choice

1. **D** Cost per ounce varies. Plug in values to form the correct equation.

2. **B** Twice a number means to multiply. Less signifies subtraction.

3. **C** More than means to add.

4. **A** Translate the text into the equation.

5. **C** To calculate how many hours, divide the distance by speed.

LESSON 6: WHAT HAPPENS WHEN . . . ?

Multiple-Choice

1. **B** Substitute values for a to see how b changes.

2. **A** Substitute values for x to see how y changes.

3. **D** When Kathy works 30 hours, she earns $240.

4. **B** Each day the amount doubles (3, 6, 12, 24, 48, 96, 192).

5. **D** Make a chart or table.

Extended Constructed Response

If you work 40 hours per week, you will make the same amount of money whether you are paid $7.50 an hour or $300.00 per week. However, if you work less than 40 hours, it is better to be paid weekly. If you are going to work more than 40 hours per week, then it is better to be paid hourly.

LESSON 7: INEQUALITIES

Multiple-Choice

1. **D** "At least" means that much money or more.

2. **B** "Between" does not include the numbers.

3. **A** Substitute values into inequality.

4. **B** Scored at least 5 goals or more.

5. **C** They need to bowl 355 or more.

CHAPTER 4

LESSON 1: MEASURES OF CENTRAL TENDENCY (MEAN, MEDIAN, AND MODE)

Multiple-Choice

1. **C** Arrange numbers in order from least to greatest. Take the average of the two middle numbers.

2. **A** $(72 + 67 + 74 + 79 + 75) \div 5$

3. **C** Total points must be 450 to have a 90 average for five tests. $450 - (87 + 94 + 82 + 88)$

4. **D** Total points must be 30 to have a mean of 5. $30 \div 6 = 5$

5. **A** The mean is already 46 after 6 games.

6. **C** $(98 + 87 + 90 + 78 + 83 + 95 + 99) \div 7$

Short Constructed Response

    7. 30   Take the average of the two middle numbers (29 and 31).

Extended Constructed Response

The average employee salary is $7.55, which is 5 cents more than what the manager reported. ($8.00 + $7.00 + $7.25 + $7.50 + $8.00 = $37.75 ÷ 5 = $7.55)

The median salary is $7.50.

LESSON 2: PRESENTING AND INTERPRETING DATA IN TABLES AND GRAPHS

Multiple-Choice

    1. **C** About 14 people chose news programs.

    2. **A** About 30 people chose reality programs.

    3. **C** $150 \cdot 40\% = 60$

    4. **B** $600 \cdot 30\% = 180$

    5. **B** $25 \cdot 20\% = 5$

    6. **A** 40% of those surveyed equals 10 people.

    7. **A** Add up all of the columns to get the total number of people.

    8. **C** There are only twice as many birthdays in April as in February.

Extended Constructed Response

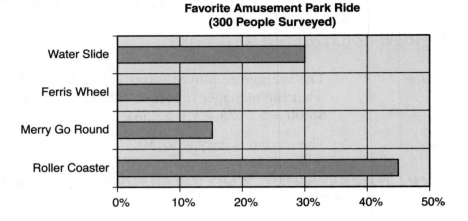

LESSON 3: DETERMINING THE PROBABILITY OF AN EVENT

Multiple-Choice

1. **B** There are 2 girls out of 5 people on the spinner.

2. **A** $\dfrac{2}{5} \cdot \dfrac{3}{5} = \dfrac{6}{25}$

3. **C** There are six numbers on each die: $6 \cdot 6 = 36$.

4. **A** $\dfrac{1}{6} \cdot \dfrac{1}{2} = \dfrac{1}{12}$

5. **D** $\dfrac{6}{15} = \dfrac{2}{5} = 40\%$

6. **B** $\dfrac{2}{3}$ of 15 marbles is 10 marbles.

Extended Constructed Response

The theoretical probability is that there will be two heads and two tails when you toss a coin four times. However, it would be possible to get all 4 heads, 3 heads and 1 tail, 2 heads and 2 tails, 1 head and 3 tails, or 4 tails.

LESSON 4: PRINCIPLES OF COUNTING (COMBINATIONS AND PERMUTATIONS)

Multiple-Choice

1. C $_3P_3$
2. B $_{12}P_3$
3. B $_8P_2$
4. A $_5C_2$
5. D $26^2 \cdot 10^2$

Short Constructed Response

6. 24 $_4P_4$

Extended Constructed Response

She can make a total of 12 different outfits.

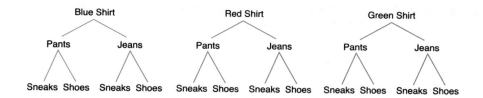

LESSON 5: VERTEX-EDGE GRAPHS AND VENN DIAGRAMS

Multiple-Choice

1. C The shortest distance is to go from *B-D-E-C*.

2. D 7 + 2 = 9

3. B There are seven paths. Do not count the direct line twice.

4. B Computer *C* can only communicate with two computers.

5. C 10 + 3 + 12 + 8 = 33

Extended Constructed Response

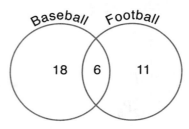

CHAPTER 5

LESSON 1: GUESS AND CHECK

1. 12 horses and 8 chickens

2. 26 quarters and 16 dimes

3. 10 quarters, 5 dimes, and 14 nickels

4. 5 cats and 12 birds

LESSON 2: MAKING A LIST, DRAWING A DIAGRAM, WORKING BACKWARDS

1. $100

2. 80

3. 4:25 P.M.

4. 24

LESSON 3: DETERMINING THE REASONABLENESS OF AN ANSWER

1. C $30 \cdot 20 = 600$

2. B Carolyn has 8 more quarters than Mark.

3. A $(3 \cdot \$20) + (2 \cdot \$10) + \$80 = \$160$

4. C $3 \cdot 24 = 72$

LESSON 4: HOW TO SOLVE A WORD PROBLEM

1. 3,200–7,200 calories

   $400 \cdot 8 = 3,200$;
   $600 \cdot 12 = 7,200$

2. 22

   Make a chart.

3. 11

   Make a list.

4. 19 feet × 12 feet = 228 square feet

   Draw a picture.

ANSWER SHEET:
PRACTICE TEST 1

Section 1
Short Constructed-Response Questions

1. $x = 4$

2. 248 cm^2

3.

4.

5.

6.

7.

8.

9.

10.

Section 2
Multiple-Choice Questions

1. Ⓐ Ⓑ Ⓒ Ⓓ   5. Ⓐ Ⓑ Ⓒ Ⓓ   9. Ⓐ Ⓑ Ⓒ Ⓓ

2. Ⓐ Ⓑ Ⓒ Ⓓ   6. Ⓐ Ⓑ Ⓒ Ⓓ   10. Ⓐ Ⓑ Ⓒ Ⓓ

3. Ⓐ Ⓑ Ⓒ Ⓓ   7. Ⓐ Ⓑ Ⓒ Ⓓ

4. Ⓐ Ⓑ Ⓒ Ⓓ   8. Ⓐ Ⓑ Ⓒ Ⓓ

Extended Constructed-Response Question

Section 3
Multiple-Choice Questions

1. Ⓐ Ⓑ Ⓒ Ⓓ   5. Ⓐ Ⓑ Ⓒ Ⓓ   9. Ⓐ Ⓑ Ⓒ Ⓓ

2. Ⓐ Ⓑ Ⓒ Ⓓ   6. Ⓐ Ⓑ Ⓒ Ⓓ   10. Ⓐ Ⓑ Ⓒ Ⓓ

3. Ⓐ Ⓑ Ⓒ Ⓓ   7. Ⓐ Ⓑ Ⓒ Ⓓ

4. Ⓐ Ⓑ Ⓒ Ⓓ   8. Ⓐ Ⓑ Ⓒ Ⓓ

Extended Constructed-Response Questions

1.

2.

Section 4
Multiple-Choice Questions

1. Ⓐ Ⓑ Ⓒ Ⓓ   5. Ⓐ Ⓑ Ⓒ Ⓓ   9. Ⓐ Ⓑ Ⓒ Ⓓ

2. Ⓐ Ⓑ Ⓒ Ⓓ   6. Ⓐ Ⓑ Ⓒ Ⓓ   10. Ⓐ Ⓑ Ⓒ Ⓓ

3. Ⓐ Ⓑ Ⓒ Ⓓ   7. Ⓐ Ⓑ Ⓒ Ⓓ   11. Ⓐ Ⓑ Ⓒ Ⓓ

4. Ⓐ Ⓑ Ⓒ Ⓓ   8. Ⓐ Ⓑ Ⓒ Ⓓ

Extended Constructed-Response Question

Section 5
Multiple-Choice Questions

1. Ⓐ Ⓑ Ⓒ Ⓓ   5. Ⓐ Ⓑ Ⓒ Ⓓ   9. Ⓐ Ⓑ Ⓒ Ⓓ

2. Ⓐ Ⓑ Ⓒ Ⓓ   6. Ⓐ Ⓑ Ⓒ Ⓓ   10. Ⓐ Ⓑ Ⓒ Ⓓ

3. Ⓐ Ⓑ Ⓒ Ⓓ   7. Ⓐ Ⓑ Ⓒ Ⓓ   11. Ⓐ Ⓑ Ⓒ Ⓓ

4. Ⓐ Ⓑ Ⓒ Ⓓ   8. Ⓐ Ⓑ Ⓒ Ⓓ

Extended Constructed-Response Question

NJ ASK PRACTICE TEST #1

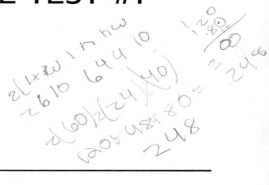

SECTION 1
SHORT CONSTRUCTED RESPONSE

Directions: For each question, write the correct answer on the line provided on the answer sheet.

1. Solve for x:

   $7x - 6 = 3x + 10$

2. What is the surface area of a rectangular prism with a length of 6 centimeters, width of 10 centimeters, and a height of 4 centimeters?

3. How many different combinations can you have when you toss five coins?

4. What is the least common multiple of 4, 6, and 18?

5. Pamela has $150.00. She goes shopping and spends $\frac{1}{2}$ of it.

   Then, she buys a book with $\frac{1}{5}$ of the money she has left.

   Finally, she takes $\frac{1}{3}$ of what she has left and puts it in the bank.

   How much money does she put in the bank?

6. Andrew participated in a bowling tournament.
   His scores for 5 games were: 178, 156, 143, 198, 186.
   What is the range of his bowling scores?

7. What is the expression for "5 more than three times a number *n*"?

8. What is the prime factorization of 260?

9. Karen, John, Sue, Tony, and Liz go to dinner. The total bill comes to $84.00, and they decide to leave a 15% tip. If they each decide to pay equally, what is the amount that each person must pay towards the total bill?

10. Which quadrilateral has four congruent sides, its opposite sides are parallel, and does NOT have four right angles?

SECTION 2
MULTIPLE-CHOICE

Directions: Darken the letter of the best answer on the answer sheet.

1. Which pair of integers has a product of –24 and a sum of +2?

    A. +12 and –10
    B. –12 and +2
    C. +6 and –4
    D. –22 and +2

2. Which characteristic is true for the polygon shown?

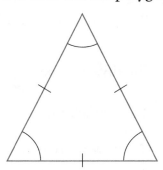

    A. Each angle measures 70°.
    B. Each side and angle is congruent.
    C. The figure has 3 angles totaling 210°.
    D. Only two sides are congruent.

3. The IZOD Center at the Meadowlands has a maximum seating capacity of 19,968 people at an NBA game. If the New Jersey Nets play 41 games there and each game is sold out, what is the best estimate for the number of fans attending the games during the regular season?

   A. 400,000

   B. 760,000

   C. 800,000

   D. 818,688

4. A kitchen table has a circular top with a diameter of 30 inches. Which statement is true about the area of the kitchen table?

   A. 94.2 in.^{2}

   B. 353.25 in.^{2}

   C. 706.5 in.^{2}

   D. 5,652 in.^{2}

5. A triangle has vertices located at X (3, –2), Y (7, –2), and Z (7, 4). If a scale factor of 3 is applied to the triangle, what are the new coordinates of vertex Y?

   A. (9, –6)

   B. (10, 1)

   C. (21, –6)

   D. (10, 7)

6. On a Saturday afternoon, you meet your friends to see a movie and get some pizza. When you return home, you have $30 in your pocket. If the movie ticket costs $8.00 and the pizza costs you $6.00, which equation could you use to determine how much money you had at the beginning of the day?

   A. $x + \$8 + \$6 = \$30$

   B. $x - \$8 - \$6 = \$30$

   C. $x + \$8 - \$6 = \$30$

   D. $x + 30 - 8 = 6$

7. Evaluate n^3 ____ p^2, when $n = 4$ and $p = 8$.

   A. $\neq$

   B. $<$

   C. $>$

   D. $=$

8. Rank the following numbers in order from GREATEST to LEAST.

$$\frac{3}{8}, \frac{2}{9}, \frac{5}{16}$$

   A. $\frac{2}{9}, \frac{5}{16}, \frac{3}{8}$

   B. $\frac{2}{9}, \frac{3}{8}, \frac{5}{16}$

   C. $\frac{5}{16}, \frac{2}{9}, \frac{3}{8}$

   D. $\frac{3}{8}, \frac{5}{16}, \frac{2}{9}$

9. What is the sum of the mode from each set of data?

   Set A: {4, 9, 7, 2, 4}

   Set S: {11, 6, 4, 6, 2}

   Set K: {−3, 2, 9, −3, −3}

   A. 4

   B. 6

   C. 7

   D. 10

7th Grade Girl's Sports (50 Students)

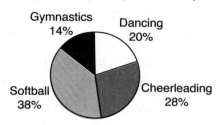

10. The above pie graph shows the results from a survey conducted about girls' sports. What can you conclude about the graph?

    A. More than 20 students chose softball.

    B. Softball was chosen by almost $\frac{2}{5}$ of those surveyed.

    C. Softball and cheerleading equal half of those surveyed.

    D. Dancing was chosen by the least number of people.

EXTENDED CONSTRUCTED RESPONSE

Directions: Write your response in the space provided on the answer sheet. Answer the questions as completely as possible.

1. Mr. McDonald has 180 feet of fencing to enclose a pen for his pigs. He will use a side of the barn, which measures 70 feet as one side of the pen. Using the fencing and the side of the barn, draw a figure and label the dimensions that will maximize the area. Explain how you obtained your answer.

SECTION 3
MULTIPLE-CHOICE

Directions: Darken the letter of the best answer on the answer sheet.

1. Three consecutive integers have a sum of 129. What is the largest of those three numbers?

   A. 42

   B. 43

   C. 44

   D. 86

2. Mr. Math's living room is shaped like a rectangle that measures 18 feet long and 15 feet wide. Ms. Language Arts's living room is also rectangular shaped, and it measures 12 feet long by 9 feet wide. Which statement is true about their living rooms?

   A. Their living rooms are neither congruent nor similar.

   B. Their living rooms are similar but not congruent.

   C. Their living rooms are congruent but not similar.

   D. Their living rooms have the same perimeter.

3. At the school dance, 50 students arrive every half hour. The dance starts at 7:00 P.M. and the last group of students arrives at 9:30 P.M. If students continually arrive every half hour, how many students are at the dance at 9:30 P.M.?

   A. 200 students

   B. 250 students

   C. 300 students

   D. 350 students

4. What is the 52nd letter in the pattern, "F A C T O R F A C T O R . . ."?

   A. A

   B. C

   C. T

   D. R

5. What is the difference between the mean of Set A and the mean of Set B?

$$\text{Set } A: \{\, 42, 17, 28, 31\}$$

$$\text{Set } B: \{\, 29, 14, 37, 8\ \}$$

   A. 29.5

   B. 22

   C. 8

   D. 7.5

6. Evaluate the expression: $3 + 2^4 - 18 \div 2 + 7$.

   A. −1

   B. 9

   C. 17

   D. 35

7. A high-definition television set is on sale for $1,820. The sale price reflects a 30% discount. What is the regular price of the television set?

   A. $6,067

   B. $2,600

   C. $1,850

   D. $1,274

8. □$MATH$ has vertices located at M (1, 4), A (1, 9), T (7, 9), H (7, 4). If the figure is translated 3 units down and 2 units to the right, which vertex would be located at (9, 1)?

   A. M

   B. A

   C. T

   D. H

9. Evaluate $3x^3 + 2y^4$ when $x = 2$ and $y = 3$.

   A. 29

   B. 42

   C. 186

   D. 1,512

10. There are 7 red marbles, 5 blue marbles, and 4 green marbles. What is the probability of selecting a red or blue marble?

    A. $\dfrac{7}{16}$

    B. $\dfrac{5}{16}$

    C. 75%

    D. 12%

EXTENDED CONSTRUCTED RESPONSE

Directions: Write your response in the space provided on the answer sheet. Answer the questions as completely as possible.

1. Determine two numbers that are greater than 72% and less than $\dfrac{19}{25}$. Write one number as a decimal and the other number as a fraction. Explain how you obtained your answer using complete sentences.

2. As part of this year's science fair, you want to see if people can correctly calculate how many seconds there are in a year that is not a leap year. Show all calculations used to determine the number of seconds in one year.

SECTION 4
MULTIPLE-CHOICE

Directions: Darken the letter of the best answer on the answer sheet.

---

1. Given the equation $4x + 3y = 36$, what happens to the value of y as x decreases?

    A. The value of y gets larger.

    B. The value of y gets smaller.

    C. The value of y must stay the same.

    D. The value of y cannot be determined from the information given.

2. Which set of data has the greatest median?

    A. 18, 12, 14, 16, 20

    B. 14, 16, 19, 15, 12

    C. 11, 14, 21, 20, 13

    D. 11, 13, 15, 17, 19

3. On a recent math test, your teacher asked you to evaluate $|x| = 4$. Of the following answer choices, which answer is best?

    A. $x = 4$

    B. $x = \pm 4$

    C. $x = (-4)$

    D. x cannot be solved based on the information given.

4. How many different combinations of outfits can you make from 4 shirts, 3 pairs of pants, and 3 pairs of shoes?

    A. 1

    B. 10

    C. 36

    D. 48

5. Which number is a square number and a cubic number?

   A. 4

   B. 16

   C. 27

   D. 64

6. A trapezoid's measurements are b_1 = 8 in., b_2 = 12 in., and its height is $\frac{1}{2}$ of the sum of its bases. What is the area of the trapezoid?

   A. 200 in.^{2}

   B. 100 in.^{2}

   C. 50 in.^{2}

   D. 5 in.^{2}

7. You are placing colored blocks in a row. The first block is red, the second block is blue, the third block is green, and the fourth block is yellow. As the pattern repeats, what colored block is in the 23rd position?

   A. Red

   B. Blue

   C. Green

   D. Yellow

8. What is the mean (average) of the employees' hourly wages from Burger Hut?

   $7.50 $8.25 $6.75 $7.00 $7.75
   $8.00 $9.50 $7.00 $8.25 $8.50

   A. $8.00

   B. $7.85

   C. $7.50

   D. $7.00

9. A particular cell phone plan charges a fee of $25.00 per month and $.15 per minute. Which equation could be used to calculate the cost of the cell phone plan each month? (Let *C* = cost and *m* = minutes.)

  A. C = $25.00 + $.15

  B. C = $25.00 + $.15*m*

  C. C = $25.15*m*

  D. C = $25.00

10. When the figure is rotated 270° clockwise, what will be its rotated image?

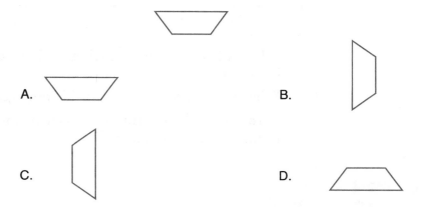

  A.

  B.

  C.

  D.

11. A bag contains 5 different colored chips. Three chips are red. Four chips are blue. Six chips are yellow. Two chips are green and one chip is purple. The probability of picking a green chip is what?

  A. Greater than picking a red chip

  B. The same as picking a blue chip

  C. Less than 13%

  D. Greater than 13%

EXTENDED CONSTRUCTED RESPONSE

Directions: Write your response in the space provided on the answer sheet. Answer the questions as completely as possible.

1. Each night in league, Tom bowls 3 games. His scores are entered in the following table.

Game 1	Game 2	Game 3	Total
186	172	209	567
226	215	183	624
156	199	187	542

■ What is his average for each game? (Round answer to the nearest hundredth.)

■ What is the median score for his nine games?

■ If Tom scored an extra 10 pins each game, what would happen to his average and median?

SECTION 5
MULTIPLE-CHOICE

Directions: Darken the letter of the best answer on the answer sheet.

1. Which of these numbers has the GREATEST number of factors?

    A. 24

    B. 36

    C. 40

    D. 61

2. Which statement is true about a right triangle?

    A. The longest side is called a hypotenuse.

    B. The Pythagorean theorem can be used to measure its sides.

    C. The triangle will always have 180 degrees.

    D. All the above statements are true.

3. The seventh grade is putting on a play to raise money for the class trip. They need to raise $600 and have decided to charge students $4.00 per ticket and adults $5.00 per ticket. Which combination of ticket sales will help the seventh grade achieve their $600 goal?

   A. 300 student tickets and 300 adult tickets

   B. 200 student tickets and 200 adult tickets

   C. 100 student tickets and 40 adult tickets

   D. 50 student tickets and 25 adult tickets

4. Examine the numbers in Set A and determine which of the following statements is true.

   Set A: { 50, 75, 80, 85, 90, 100 }

   A. The mean of Set A is 80, and the median is 80.

   B. The mean of Set A is 80, and the mode is 82.5.

   C. The mean of Set A is 80, and the range is 50.

   D. The mean, median, and mode of Set A are 80.

5. Three consecutive integers have a sum of 108. Which number is the smallest?

   A. 35

   B. 36

   C. 37

   D. 106

6. A shipping container is shaped like a rectangular prism. The base has a length of 14 inches and a width of 8 inches. In order to have a volume of 672 cubic inches, what is the height of the container?

   A. 48 inches

   B. 24 inches

   C. 12 inches

   D. 6 inches

7. Point J is located at $(-5, -3)$. What transformation must be performed to have Point J' be located at $(5, 3)$?

   A. Translate the point up 10 units and right 6 units.

   B. Reflect the point over the x-axis and then reflect it over the y-axis.

   C. Rotate the point 180 degrees.

   D. Move the point diagonally 5 units.

8. Shenice has three shirts, four pairs of pants, and three pairs of shoes. How many different outfits can Shenice create?

   A. 36

   B. 24

   C. 12

   D. 10

9. Which value of x will NOT satisfy the following inequality?

$$3(x - 12) > -24$$

   A. 10

   B. 8

   C. 6

   D. 4

10. What is the probability that a number rolled on a cube will NOT be a factor of 6?

   A. $\dfrac{2}{3}$

   B. $\dfrac{1}{2}$

   C. $\dfrac{1}{3}$

   D. $\dfrac{1}{6}$

11. What is the difference between the volume of a cube measuring 5 inches per side and a cube measuring 3 inches per side?

    A. 125 cubic inches

    B. 98 cubic inches

    C. 27 cubic inches

    D. 8 cubic inches

EXTENDED CONSTRUCTED RESPONSE

Directions: Write your response in the space provided on the answer sheet. Answer the questions as completely as possible.

1. Find all numbers that satisfy the following statements:

- I am an integer greater than 0 and less than 100.

- I am a multiple of 3.

- The sum of my digits is a multiple of 9.

Show all work and explain how you obtained your answer.

ANSWER SHEET:
PRACTICE TEST 2

Section 1
Short Constructed-Response Questions

1. _____

2. _____

3. _____

4. _____

5. _____

6. _____

7. _____

8. _____

9. _____

10. _____

Section 2
Multiple-Choice Questions

1. Ⓐ Ⓑ Ⓒ Ⓓ   5. Ⓐ Ⓑ Ⓒ Ⓓ   9. Ⓐ Ⓑ Ⓒ Ⓓ

2. Ⓐ Ⓑ Ⓒ Ⓓ   6. Ⓐ Ⓑ Ⓒ Ⓓ   10. Ⓐ Ⓑ Ⓒ Ⓓ

3. Ⓐ Ⓑ Ⓒ Ⓓ   7. Ⓐ Ⓑ Ⓒ Ⓓ

4. Ⓐ Ⓑ Ⓒ Ⓓ   8. Ⓐ Ⓑ Ⓒ Ⓓ

Extended Constructed-Response Question

Section 3
Multiple-Choice Questions

1. Ⓐ Ⓑ Ⓒ Ⓓ   5. Ⓐ Ⓑ Ⓒ Ⓓ   9. Ⓐ Ⓑ Ⓒ Ⓓ

2. Ⓐ Ⓑ Ⓒ Ⓓ   6. Ⓐ Ⓑ Ⓒ Ⓓ   10. Ⓐ Ⓑ Ⓒ Ⓓ

3. Ⓐ Ⓑ Ⓒ Ⓓ   7. Ⓐ Ⓑ Ⓒ Ⓓ

4. Ⓐ Ⓑ Ⓒ Ⓓ   8. Ⓐ Ⓑ Ⓒ Ⓓ

Extended Constructed-Response Questions

1.

2.

Section 4
Multiple-Choice Questions

1. Ⓐ Ⓑ Ⓒ Ⓓ   5. Ⓐ Ⓑ Ⓒ Ⓓ   9. Ⓐ Ⓑ Ⓒ Ⓓ

2. Ⓐ Ⓑ Ⓒ Ⓓ   6. Ⓐ Ⓑ Ⓒ Ⓓ   10. Ⓐ Ⓑ Ⓒ Ⓓ

3. Ⓐ Ⓑ Ⓒ Ⓓ   7. Ⓐ Ⓑ Ⓒ Ⓓ   11. Ⓐ Ⓑ Ⓒ Ⓓ

4. Ⓐ Ⓑ Ⓒ Ⓓ   8. Ⓐ Ⓑ Ⓒ Ⓓ

Extended Constructed-Response Question

Section 5
Multiple-Choice Questions

1. Ⓐ Ⓑ Ⓒ Ⓓ   5. Ⓐ Ⓑ Ⓒ Ⓓ   9. Ⓐ Ⓑ Ⓒ Ⓓ

2. Ⓐ Ⓑ Ⓒ Ⓓ   6. Ⓐ Ⓑ Ⓒ Ⓓ   10. Ⓐ Ⓑ Ⓒ Ⓓ

3. Ⓐ Ⓑ Ⓒ Ⓓ   7. Ⓐ Ⓑ Ⓒ Ⓓ   11. Ⓐ Ⓑ Ⓒ Ⓓ

4. Ⓐ Ⓑ Ⓒ Ⓓ   8. Ⓐ Ⓑ Ⓒ Ⓓ

Extended Constructed-Response Question

NJ ASK PRACTICE TEST #2

SECTION 1
SHORT CONSTRUCTED RESPONSE

Directions: For each question, write the correct answer on the line provided on the answer sheet.

1. Solve for x:

$$3x + 14 + 5x = 2x + 62$$

2. What is the height of a trapezoid whose area is 99 in.^{2}, $b_1 = 8$ in., and $b_2 = 14$ in.?

3. What is the probability of tossing a coin and getting "heads" and rolling a die and getting "an even number"?

4. How many millimeters are there in $2\frac{3}{4}$ km?

5. How many different ways can you spell TIME by moving downward and diagonally with each row?

T

I   I

M M M

E   E   E   E

6. There is a 70% chance of rain on Saturday and a 50% chance of rain on Sunday. What is the chance that it will rain on both days?

7. Kimi uses 450 minutes a month on her cell phone. What is a reasonable estimate for the number of minutes she uses her cell phone in one year?

8. Solve for x: $6x + 15 = 4x + 19.5$.

9. A full box of crackers contains 8 servings and 150 calories per serving. How many calories are there in an entire box?

10. A circle has a circumference of 53.38 cm. What is the radius of the circle?

SECTION 2
MULTIPLE-CHOICE

Directions: Darken the letter of the best answer on the answer sheet.

1. What statement is true about the following set of test scores?

    {89, 75, 90, 94, 82, 86, 88, 71, 94, 76}

   A. The mean is 84.5.

   B. The median is 87.

   C. The mode is 94.

   D. All the above statements are true.

2. Which graph best expresses the solution to the inequality $x + 5 < 8$?

A.

B.
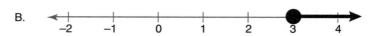

C.

D.

3. There are 80 students in the school chorus. At yesterday's rehearsal, the choir director mentioned that 30% of the chorus was absent. How many students attended the rehearsal?

    A. 56

    B. 30

    C. 24

    D. 3

4. How many millimeters are there in $2\frac{1}{2}$ km ?

    A. 2,500,000 mm

    B. 250,000 mm

    C. 25,000 mm

    D. 2,500 mm

5. Charlie went to a local fast food restaurant and ordered a Super Drink that contained 48 ounces. If he drank 2 Super Drinks in one day, how many cups did he drink?

    A. 24 cups

    B. 18 cups

    C. 12 cups

    D. 6 cups

6. Examine the results from a survey about favorite football teams.

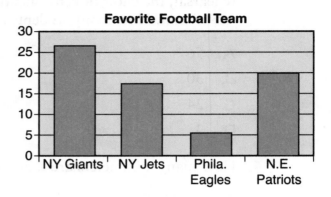

Favorite Football Team

Which statement is FALSE?

A. About half as many people like the NY Giants as like the NY Jets.

B. About three times as many people like the NY Giants than the Philadelphia Eagles.

C. More people like the New England Patriots than the NY Jets.

D. The number of people who like the Philadelphia Eagles and the New England Patriots is equal to the number of people who like the NY Giants.

7. The $\sqrt{498}$ is between what two consecutive integers?

A. 20 and 21

B. 21 and 22

C. 22 and 23

D. 22 and 24

8. Which statement is TRUE about the expression $4^3 + 2^2$?

A. It has the same value as 6^6.

B. It has a value greater than 9^2.

C. It has a value less than 2^6.

D. It has the same value as $10^2 - 2^5$.

9. Five athletes are chosen to represent their sports at the school pep rally. Two girls and three boys are selected to walk into the gymnasium. If a girl must enter first and the other girl must be the last one to enter, how many different ways can the athletes enter the gym?

    A. 120

    B. 60

    C. 12

    D. 6

10. Your friend sends you a present in the mail. When the package arrives, the directions state to find its dimensions and guess what is inside. If the package has a length of 8 inches, width of 6 inches, and a height of 6 inches, what is the package's volume?

    A. 20 in.^{3}

    B. 48 in.^{3}

    C. 216 in.^{3}

    D. 288 in.^{3}

EXTENDED CONSTRUCTED RESPONSE

Directions: Write your response in the space provided on the answer sheet. Answer the questions as completely as possible.

1. Melinda is currently training for a dance competition. She realizes that she is out of shape and needs to work on her endurance. Her coach and instructor advise her to walk for 5 minutes per day for the first week and double the amount each week thereafter.

■ Construct a chart or table showing how much she walks per day during the first five (5) weeks of her training.

■ Draw a graph showing the amount of time she trains each week.

■ If she continues, how many minutes she will walk per day during the eighth week, and tell whether or not that amount is reasonable?

SECTION 3
MULTIPLE-CHOICE

Directions: Darken the letter of the best answer on the answer sheet.

1. What is the reflected image of the following figure over a horizontal line?

A.

B.

C.

D.

Practice Test #2 ▪ 231

2. Seven less than three times a number is twenty-nine. What is the number?

   A. 12
   B. 10
   C. 29
   D. 36

3. Every week John saves $8.00 to his bank account. Currently, he has $25.00 in his bank account. If he saves his money for w weeks, which expression could be used to determine how much money he will have in his bank account?

   A. $8.00 + 25.00w$
   B. $25.00 + 8.00w$
   C. $8.00 + w + $25.00
   D. ($25.00 + $8.00)$w$

4. There is a 20% chance that it will snow on Tuesday and a 50% chance that it will snow on Wednesday. What is the probability that it will snow on either day?

   A. 10%
   B. 20%
   C. 50%
   D. 70%

$0.2 \times 0.5 = 1.0 = 10\%$

5. The m∠MKN = 81°. Its complement measures how many degrees?

   A. 89°
   B. 81°
   C. 19°
   D. 9°

6. The Fibonacci Sequence is 1, 1, 2, 3, 5, 8, ... As the sequence continues, what would be the 10th number?

    A. 34

    B. 55

    C. 89

    D. 144

7. Which of the following numbers is divisible by 2, 3, and 8?

    A. 84

    B. 336

    C. 490

    D. 800

8. If a coin is tossed 24 times, what is the theoretical probability of the event?

    A. There will be an unequal amount of heads and tails.

    B. There will be an equal amount of heads and tails.

    C. There will be 24 heads.

    D. There will be 24 tails.

9. Blair has 7 red scarves and 3 blue scarves in a drawer. What is the probability that she could randomly choose 2 red scarves?

    A. $\dfrac{6}{100}$

    B. $\dfrac{7}{10}$

    C. $\dfrac{21}{50}$

    D. $\dfrac{7}{15}$

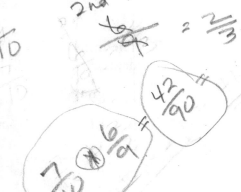

10. You need to find a number between $\frac{3}{5}$ and $\frac{7}{9}$. Which number is NOT between $\frac{3}{5}$ and $\frac{7}{9}$?

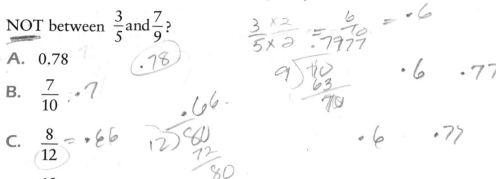

A. 0.78

B. $\frac{7}{10}$

C. $\frac{8}{12}$

D. $\frac{13}{21}$

EXTENDED CONSTRUCTED RESPONSE

Directions: Write your response in the space provided on the answer sheet. Answer the questions as completely as possible.

1. Your school is having a fundraiser and will be selling school sweatshirts with the proceeds benefiting the Student Council. ASK Clothing regularly sells their sweatshirts for $19.95 but is willing to give you a 20% discount per sweatshirt. Their competitor, TFN Apparel, normally sells their sweatshirts for $17.95 and is willing to give you a 10% discount. Which store has the better bargain? Show your calculations and explain your reasoning using complete sentences. If necessary, round your answers to the nearest penny.

2. Lucas and Jamal are downloading songs to their media player. On the first day, they download 1 song. They like it so much that they download 2 songs on the next day. On the third day, they download 4 songs. If the pattern continues how many songs will they download on the 12th day and how many songs will they have downloaded in all? Show your work and explain how you obtained your answer.

SECTION 4
MULTIPLE-CHOICE

Directions: Darken the letter of the best answer on the answer sheet.

1.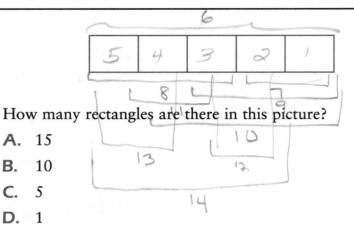

   How many rectangles are there in this picture?

   A. 15

   B. 10

   C. 5

   D. 1

2. When baking cupcakes, Shelley uses $\frac{3}{4}$ cup of walnuts per batch.

   Each batch yields 18 cupcakes. If Shelley is baking cupcakes for the entire class of 162 students, which proportion could Shelley use to determine how many cups of walnuts she will need?

   A. $\dfrac{\frac{3}{4}}{18} = \dfrac{162}{x}$

   B. $\dfrac{162}{18} = \dfrac{\frac{3}{4}}{x}$

   C. $\dfrac{162}{\frac{3}{4}} = \dfrac{x}{18}$

   D. $\dfrac{18}{\frac{3}{4}} = \dfrac{162}{x}$

3. If you double a rectangle's length and width, what changes occur with the rectangle's perimeter?

   A. The rectangle's perimeter remains the same.

   B. The rectangle's perimeter doubles.

   C. The rectangle's perimeter quadruples.

   D. The rectangle's perimeter is multiplied by 8.

4. Which equation can be expressed as a number n divided by 3 increased by 7 is 11?

   A. $\frac{n}{3} + 7 = 11$

   B. $3n + 7 = 11$

   C. $\frac{n}{7} + 3 = 11$

   D. $(n \div 7 + 3) = 11$

5. Which function is modeled by the table?

X	Y
1	−1
2	0
3	1
4	2
5	3

   A. $Y = 2X - 4$

   B. $Y = 3X - 5$

   C. $Y = -X - 2$

   D. $Y = X - 2$

6.

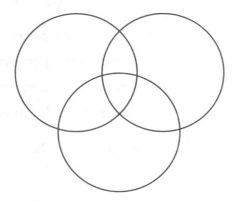

How many regions are formed by these three interlocking circles?

A. 3

B. 5

C. 7

D. 8

7. Which number has the greatest number of different prime factors?

A. 195

B. 169

C. 142

D. 98

8.

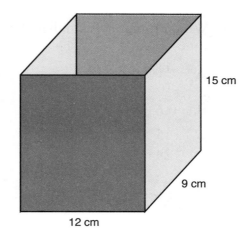

15 cm

9 cm

12 cm

The box shown does not have a top. What is the surface area of the box if its length measures 12 cm, its width is 9 cm, and it has a height of 15 cm?

A. 846 in.^{2}

B. 738 in.^{2}

C. 711 in.^{2}

D. 486 in.^{2}

9. What number can be expressed as $2^3 \cdot 3 \cdot 5^2$?

A. 180

B. 240

C. 300

D. 600

10. Based on estimates, the town of Wayne has a population of 54,000 people. Wayne has an area of approximately 25 square miles. What is the population density of Wayne?

A. 2,160 people per square mile

B. 2,000 people per square mile

C. 2,500 people per square mile

D. 5,000 people per square mile

11. Which number is smaller than $\frac{5}{8}$?

    A. $\frac{11}{16}$

    B. 0.6

    C. $\frac{15}{24}$

    D. $\frac{3}{4}$

EXTENDED CONSTRUCTED RESPONSE

Directions: Write your response in the space provided on the answer sheet. Answer the questions as completely as possible.

1. Your seventh grade class has conducted a survey to find out what is the most popular flavor of ice cream. The following information was gathered.

    Vanilla—28%
    Chocolate—40%
    Strawberry—12%
    Chocolate Chip Mint—15%
    Other—5%

■ Create a graph to display the results.

■ Describe why you chose the graph and why it best represents the data.

■ Express each percentage as a fraction.

SECTION 5
MULTIPLE-CHOICE

Directions: Darken the letter of the best answer on the answer sheet.

---

1. If a person standing 5 feet tall casts a shadow 12 feet long, then how long is a tree's shadow if a tree stands 17.5 feet tall?

    A. 42 feet

    B. 24.5 feet

    C. 17 feet

    D. 12 feet

2. The picture below shows a magic square. In a magic square, each row, column, and diagonal have the same sum. The numbers in the magic square are the integers from 1 to 9 inclusive. What is the value of C?

8	1	A
3	5	B
4	9	C

    A. 15

    B. 7

    C. 6

    D. 2

3. What is the 40th digit in the pattern $0.\overline{142857}$?

    A. 1

    B. 2

    C. 4

    D. 8

4. Mark and Juanita want to tile their kitchen floor. If the kitchen floor is shaped like a rectangle and measures 11 feet × 14 feet, how many square tiles will they need to cover the floor if each square tile measures 4 inches × 4 inches?

   A. 1,386

   B. 154

   C. 16

   D. 9.625

5. What is the area of the shaded region in this diagram?

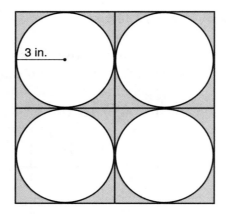

   A. 113.04 in.^{2}

   B. 30.96 in.^{2}

   C. 28.26 in.^{2}

   D. 12 in.^{2}

6. A license plate is made up of three letters and three digits. How many different combinations of letters and digits can be created using three letters and three digits?

   A. $26^3 \cdot 10^3$

   B. 26^3

   C. 10^3

   D. $26 \cdot 10$

7. At a local restaurant, you order a cheeseburger that costs $3.75, an order of french fries that costs $2.95, and a small drink that costs $1.30. There is 7% sales tax added to the order. What is the total cost of your meal?

 A. $ .56

 B. $8.00

 C. $8.56

 D. $8.63

8. If $3x + 7 = 34$, what is the value of $5x$?

 A. 9

 B. 27

 C. 45

 D. 59

9. Point R is located at $(-3, 4)$ and the point is translated +4 units horizontally. Where will the translated point be located?

 A. On the x-axis

 B. On the y-axis

 C. In quadrant I

 D. In quadrant II

10. Examine the following set of numbers. {1, 3, 6, 10, . . .} If the pattern continues, what number would be the sixth term?

 A. 28

 B. 21

 C. 15

 D. 13

11. The following sets of numbers are listed.

Set *A* {2, 5, 8, 11, 14}

Set *B* {3, 6, 9, 12, 15}

What is the mean of Set *A* + the median of Set *B*?

A. 29

B. 17

C. 15

D. 8

EXTENDED CONSTRUCTED RESPONSE

Directions: Write your response in the space provided on the answer sheet. Answer the questions as completely as possible.

1. A special lottery game is being played where a player must guess a three-digit number. The first digit must be an odd number. The second digit must be an even number, and the third number must be a prime digit. How many different combinations are possible and what is the likelihood that the entire number is even? Show your work and explain how you obtained your answer.

ANSWERS TO PRACTICE TESTS

PRACTICE TEST #1

SECTION 1—SHORT CONSTRUCTED RESPONSE

1.   4 $7x - 6 = 3x + 10$
$$4x = 16$$
$$x = 4$$

2.   248 cm² $SA = 2lw + 2wh + 2lh$
$$= (2 \cdot 6 \cdot 10) + (2 \cdot 10 \cdot 4) + (2 \cdot 6 \cdot 4)$$
$$= 248 \, \text{cm}^2$$

3.   32   There are two different outcomes on each coin: $2 \cdot 2 \cdot 2 \cdot 2 \cdot 2 = 32$.

4.   36   36 is the smallest multiple of 4, 6, and 18.

5.   \$20.00 $150 \cdot \dfrac{1}{2} = 75$
$$75 \cdot \dfrac{1}{5} = 15$$
$$75 - 15 = 60$$
$$60 \cdot \dfrac{1}{3} = 20$$

6.   55   Range = Highest score – Lowest score
198 – 143 = 55

7. $3n + 5$

8.   260 $2 \cdot 2 \cdot 5 \cdot 13$

9. **$19.32**   Bill = $84.00, tip = 15% of $84.00 = $12.60, total bill = $96.60; each person pays $96.00 ÷ 5 = $19.32

10. Rhombus

SECTION 2—MULTIPLE-CHOICE

1. **C** $+6 \cdot (-4) = -24, +6 + (-4) = +2$

2. **B**   In an equilateral triangle, all sides and angles are congruent as based on the markings.

3. **C** $20,000 \cdot 40 = 800,000$

4. **C** $A = \pi r^2$; $A = (3.14)(15^2) = 706.5$ in.^{2}

5. **C** $(7 \cdot 3, (-2) \cdot 3)$; $(21, -6)$

6. **B**   Your original amount − $8 − $6 = $30.

7. **D** $4^3 = 8^2$; $64 = 64$

8. **D**   As decimals: $\frac{3}{8} = 0.375, \frac{5}{16} = 0.3125, \frac{2}{9} = 0.\overline{2}$

9. **C**   Mode for set $A = 4$, set $B = 6$, set $C = (-3)$; $4 + 6 + (-3) = 7$.

10. **B**   38% is close to 40%, which is $\frac{2}{5}$.

EXTENDED CONSTRUCTED RESPONSE

1. In order to create the most area for his farm while using the side of the barn, Mr. McDonald will use the fencing to create a square. He will divide the 180 feet of fencing to create three sides each measuring 60 feet. This will create a maximum area of 3,600 square feet. Note that you do not have to use the entire side of the barn.

SECTION 3—MULTIPLE-CHOICE

1. **C** $42 + 43 + 44 = 129$

2. **A**   The living rooms are not congruent because they are not the same size and shape. The living

rooms are not similar because their dimensions
are not proportional.

3. **C** There are six half-hour intervals, so $6 \cdot 50 = 300$.

4. **C** $52 \div 6 = 8r4$. The fourth letter in the series is T.

5. **D** Mean of set $A = 29.5$. Mean of set $B = 22$.
$29.5 - 22 = 7.5$.

6. **C** Follow the order of operations.

7. **B** $\dfrac{\$1,820}{x} = \dfrac{70}{100}$; $x = \$2,600$

8. **D** Add $+2$ units to the x-coordinate and (-3) units
to the y-coordinate.

9. **C** Substitute and follow the order of operations:
$3(2^3) + 2(3^4) = 186$.

10. **C** There are 12 marbles that are either red or blue
out of a total of 16 marbles. Therefore,
$\dfrac{12}{16} = 75\%$.

EXTENDED CONSTRUCTED RESPONSE

1. Both numbers must be between $\dfrac{72}{100}$ and $\dfrac{76}{100}$. One
number must be written as a decimal and the other as
a fraction.

2. 365 days × 24 hours/day = 8,760 hours

8,760 hours × 60 minutes/hour = 525,600 minutes

525,600 minutes × 60 seconds/minute = 31,536,000 seconds

SECTION 4—MULTIPLE-CHOICE

1. **A** Substitute values of x to see how y changes.

2. **A** Arrange each set in order from least to greatest
and evaluate the middle numbers within each
answer choice.

3. **B** $+4$ and (-4) are each 4 units from zero.

4. **C** $4 \cdot 3 \cdot 3 = 36$

5. D $8^2 = 64$; $4^3 = 64$

6. B $A = \dfrac{1}{2}h(b_1 + b_2)$; $A = \dfrac{1}{2}(10)(8 + 12)$; $A = 100$ in.^{2}

7. C $23 \div 4 = 5r3$; the third color in the sequence is green.

8. B   Add up the salaries and divide by 10: $78.50 \div 10 = \$7.85$.

9. B   Monthly fee is fixed. The minutes charged are varied each month.

10. B   Turn figure $\dfrac{3}{4}$.

11. C   The probability of picking a green chip is 2 out of 16 or 12.5%.

EXTENDED CONSTRUCTED RESPONSE

▪ Tom's average is 192.56.

▪ Tom's median score is 187.

▪ If Tom scored an extra ten pins per game, his average and his median would each increase by 10.

SECTION 5—MULTIPLE-CHOICE

1. B   There are 9 factors of 36 (1, 2, 3, 4, 6, 9, 12, 18, 36).

2. D   All the statements are true.

3. C $(100 \cdot \$4 + 40 \cdot \$5) = \$600$

4. C   Mean = 80 because $(50 + 75 + 80 + 85 + 90 + 100) \div 6 = 80$; range = $100 - 50 = 50.125$ in.^{3} − 27in.^{3} = 98 in.^{3}

5. A $35 + 36 + 37 = 108$

6. D   14 in. · 8 in. · 6 in. = 672 in.^{3}

7. B   A reflection over both axes when the point is in quadrant III will move the point to quadrant I.

8. A $3 \cdot 4 \cdot 3 = 36$

9. D $x = 4$; x cannot equal 4 as it must be greater than 4 to satisfy the inequality.

10. C   4 and 5 are not factors of 6.

11. B 125 in.^{3} − 27 in.^{3} = 98 in.^{3}

EXTENDED CONSTRUCTED RESPONSE

1. The numbers that satisfy the statements are 9, 18, 27, 36, 45, 54, 63, 72, 81, 90, and 99. To obtain my answer, I made a list of all integers that are multiples of 3 greater than 0 and less than 100. Then, I found those numbers whose digits have a sum that is a multiple of 9.

PRACTICE TEST #2

SECTION 1—SHORT CONSTRUCTED RESPONSE

1. 8
$$3x + 14 + 5x = 2x + 62$$
$$8x + 14 = 2x + 62$$
$$6x = 48$$
$$x = 8$$

2. 9 in.
$$A = \frac{1}{2}h(b_1 + b_2)$$
$$99 = \frac{1}{2}h(8 + 14)$$
$$99 = \frac{1}{2}h(22)$$
$$99 = 11h$$
$$9 = h$$

3. $\dfrac{1}{4}$ $\dfrac{1}{2} \cdot \dfrac{1}{2} = \dfrac{1}{4}$

4. 2,750 mm $2\dfrac{3}{4} \cdot 1,000 = 2,750$

5. 8   Follow the paths.

6. 35% $0.7 \cdot 0.5 = 0.35$; $0.35 = 35\%$

7. 6,000 $500 \cdot 12 = 6,000$

8. 2.25 $6x + 15 = 4x + 19.5$
$$2x = 4.5$$
$$x = 2.25$$

9. 1,200 $8 \cdot 150 = 1,200$

10. 8.5cm
$$C = 2\pi r$$
$$53.38 = 2(3.14)r$$
$$53.38 = 6.28r$$
$$53.38 \div 6.28 = 6.28r \div 6.28$$
$$8.5 = r$$

SECTION 2—MULTIPLE-CHOICE

1. **D**   Calculate all answers. Every measure is true.

2. **D**   Answer is not equal to 3, thus open circle. Shade to left since it is less than.

3. **A**   70% are present: $80 \cdot 70\% = 56$.

4. **A**   There are 100,000 mm in 1 km.

5. **C** $96 \div 8 = 12$

6. **A**   More people like the Giants than the Jets.

7. **C**   Use the calculator to determine the square root.

8. **D** $4^3 + 2^2 = 64 + 4 = 68$

9. **C** $2 \cdot 3 \cdot 2 \cdot 1 \cdot 1 = 12$

10. **D** $V = lwh = 8 \cdot 6 \cdot 6 = 288$

EXTENDED CONSTRUCTED RESPONSE

1.

Week	Time
1	5
2	10
3	20
4	40
5	80

Training Schedule

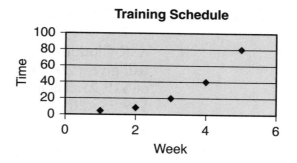

Melinda will walk 640 minutes a day during the eighth week, and that amount is not reasonable.

SECTION 3—MULTIPLE-CHOICE

1. **B** Flip the drawing upside down.

2. **A** $3n - 7 = 29$; $3n = 36$; $n = 12$

3. **B** $8.00w$ represents how much money John saves per week.

4. **A** Convert percents to decimals and multiply. $0.2 \cdot 0.5 = 0.1$

5. **D** $90 - 81 = 9$

6. **B** Add the previous two numbers to get the next number in the sequence.

7. **B** Use divisibility rules to test answer choices.

8. **B** In theory, there are two different outcomes and each has an equally likely chance of appearing.

9. **D** $\dfrac{7}{10} \cdot \dfrac{6}{9} = \dfrac{42}{90} = \dfrac{7}{15}$

10. **A** Convert fractions to decimals and test answer choices.

EXTENDED CONSTRUCTED RESPONSE

1. ASK Clothing has the better sale price. ASK Clothing will give a 20% discount off of a regular price of $19.95. Their sale price is determined by $19.95 · 20% = $3.99; $19.95 − $3.99 = $15.96.

    TFN Apparel will give a 10% discount off of a regular price of $17.95. Their sale price is determined by $17.95 · 10% = $1.795; $17.95 − $1.795 = $16.155 ≈ $16.16.

    Finally, ASK Clothing will sell the sweatshirts for $15.96 and TFN Apparel will sell the sweatshirts for $16.16.

2. Make a table to see how many songs are downloaded on the 12th day. The amount of songs double each day.

Day	1	2	3	4	5	6	7	8
Songs	1	2	4	8	16	32	64	128

Lucas and Jamal will download 128 songs on the 8th day and 255 songs in total.

SECTION 4—MULTIPLE-CHOICE

1. **A** 5 + 4 + 3 + 2 + 1 = 15

2. **D** Make sure proportion is set up correctly.

3. **B** Substitute numbers and test results.

4. **A** Translate text into equation.

5. **D** Substitute values of X into each equation and determine the correct output for Y.

6. **C** Count the regions of the circles in the Venn diagram.

7. **A** Find the prime factorization for each answer choice.

8. **B**   SA = $lw + 2wh + 2lh$ = $(12 \cdot 9) + (2 \cdot 9 \cdot 15) +$ $(2 \cdot 12 \cdot 15)$

9. **D** $2^3 \cdot 3 \cdot 5^2 = 8 \cdot 3 \cdot 25 = 600$

10. **A** $54,000 \div 25 = 2,160$

11. **B**   Convert fraction to decimal. 0.6 is smaller.

EXTENDED CONSTRUCTED RESPONSE

1.   A bar graph would be the most appropriate to show the comparison between the flavors.

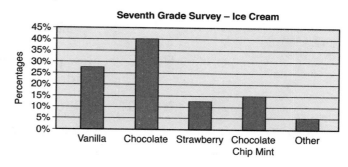

Vanilla: $28\% = \dfrac{7}{25}$;   Chocolate: $40\% = \dfrac{2}{5}$;

Strawberry: $12\% = \dfrac{3}{25}$;   Chocolate Chip Mint: $15\% = \dfrac{3}{20}$;

Other: $5\% = \dfrac{1}{20}$

SECTION 5—MULTIPLE-CHOICE

1.   **A** $\dfrac{5}{12} = \dfrac{17.5}{x}$
     $5x = 12(17.5)$
     $5x = 210$
     $x = 42$

2.   **D**   Each row, column, and diagonal has a sum of 15. $4 + 9 + C = 15$.

3.   **D**   Count the digits or $40 \div 6 = 6r4$.

4. A $11\,\text{ft} \cdot 14\,\text{ft} = 154\,\text{ft}^2$
        $154\,\text{ft}^2 \cdot 144 = 22{,}176\,\text{in}^2$
        $22{,}176 \div 16 = 1{,}386$

5. B $144 - 4(9)(3.14) = 30.96$

6. A $26 \cdot 26 \cdot 26 \cdot 10 \cdot 10 \cdot 10$

7. C $(\$3.75 + \$2.95 + \$1.30) \cdot 7\% = \$.56$
        $\$8.00 + \$.56 = \$8.56$

8. C $3x + 7 = 34$
        $3x = 27$
        $x = 9$
        $5x = 5(9) = 45$

9. C   New point is located at (1, 4) which is found in quadrant I.

10. B   1, 3, 6, 10, 15, **21**

11. B   The mean of set A = 8. The median of set B is 9. $8 + 9 = 17$.

EXTENDED CONSTRUCTED RESPONSE

1. There are five possible numbers for the first digit and five possible numbers for the second digit. However, there are four prime digits (2, 3, 5, 7). There are 100 different numbers possible because $5 \cdot 5 \cdot 4 = 100$. Since there is only one even prime number and the prime number will be the last digit, we can calculate the number of three-digit even combinations to be $5 \cdot 5 \cdot 1 = 25$. Finally, there is a $\dfrac{25}{100} = \dfrac{1}{4}$ chance of getting an even number: $\dfrac{25}{100} = \dfrac{1}{4}$.

INDEX